INTRODUCTION TO THE THEORY OF PARTIAL DIFFERENTIAL EQUATIONS

The New University Mathematics Series

Editor:
Professor E. T. DAVIES
Department of Mathematics, University of Southampton

This series is intended for readers whose main interest is in mathematics, or who need the methods of mathematics in the study of science and technology. Some of the books will provide a sound treatment of topics essential in any mathematical training, while other, more advanced, volumes will be suitable as preliminary reading for research in the field covered. New titles will be added from time to time.

BONSALL, BROWN and PAGE: *Elements of Functional Analysis*
BURGESS: *Analytical Topology*
COOPER: *Functions of Real Variables*
CURLE and DAVIES: *Modern Fluid Dynamics*
MAUNDER: *Introduction to Algebraic Topology*
PORTEOUS: *Topological Geometry*
RUND: *The Hamilton-Jacobi Theory in the Calculus of Variations*
SAMET: *Numerical Analysis*
SMITH: *Laplace Transform Theory*
SMITH: *Introduction to the Theory of Partial Differential Equations*
SPAIN: *Vector Analysis*
ZERMANSKY: *Linear Algebra and Analysis*

Introduction to the

Theory of Partial Differential Equations

M. G. SMITH

Principal Lecturer, Department of Mathematics
Sir John Cass College, London

D. VAN NOSTRAND COMPANY LTD
LONDON

PRINCETON, NEW JERSEY TORONTO

D. VAN NOSTRAND COMPANY LTD.
Windsor House, 46 Victoria Street, London, SW1

D. VAN NOSTRAND COMPANY INC.
Princeton, New Jersey

VAN NOSTRAND REGIONAL OFFICES:
New York, Chicago, San Francisco

D. VAN NOSTRAND COMPANY (CANADA) LTD.
Toronto

Published simultaneously in Canada by
D. VAN NOSTRAND COMPANY (Canada), LTD.

Library of Congress Catalog Card No. 67–19863

PRINTED IN GREAT BRITAIN BY
BUTLER & TANNER LTD, FROME AND LONDON

Preface

Too frequently, undergraduates taking courses for an honours degree in mathematics first encounter partial differential equations in the mathematical physics part of their curricula, when solutions to the elementary problems that occur are conjured *ad hoc* from nowhere, as if by magic. Thus I have found undergraduates quite surprised to learn that a systematic approach might be possible. I suspect that students of physics and theoretical chemistry or engineering are treated in much the same way, although in all probability the problems which are considered will be of greater physical interest than those the budding mathematician is presented with.

I have therefore attempted to produce an introduction to the theory of partial differential equations, which I hope is systematic. It should be suitable for the final year of an honours mathematics course, or for a course for an M.Sc. by examination, which is now becoming common in British universities. I also hope that it could be of profit to scientists of other faculties whose interests are theoretical. I have therefore assumed that the reader has some previous knowledge of the theory of ordinary differential equations, of some complex analysis, and is familiar with the more commonly occurring functions of mathematical physics. I have not assumed that he is aware of the more recondite properties of these functions.

It will be objected that this approach is a classical one. This is freely admitted. The main reason for this is that it is intended first and foremost for the mathematical physicist rather than the pure mathematician, even though it is analytical. It should also provide an introduction to books which will give the more modern approach. In particular the student is recommended to the volumes by COURANT [1], GARABEDIAN [4] and HÖRMANDER [9].

I would like to acknowledge, with great thanks, the help I have received from a number of people: those who read the book in typescript and suggested a number of additions and improvements, in particular Mr C. W. Celia of Sir John Cass College; my wife, who, as usual, bore the brunt of preparing the typescript, drawing the diagrams and reading my manuscript; the Editorial department of D. Van

Nostrand Company Ltd, whom I have always found most helpful, and their referee who made a number of valuable comments and suggestions; lastly, but certainly not least, Dr J. W. Maccoll, who first raised my interest in this subject, and who persuaded me of the pleasure of reading Hadamard. If this volume has any moment, it will be primarily due to his influence.

Sir John Cass College M. G. Smith
January 1967

Contents

CHAPTER 1

The Partial Differential Equations of Mathematical Physics

1.1 Introduction

Virtually every problem that can be formulated in mathematical physics involves the solution of a partial differential equation. There are problems which require only the solution of ordinary differential equations, but these are usually obtained by introducing simplifying assumptions into a more general problem governed by a partial differential equation. In any case, if the solution of the problem is made to depend on an ordinary differential equation, whether directly or as a stage in the solution of a partial differential equation, the problem can be regarded as solved. This is because the theory of ordinary equations is so well developed and provided the boundary conditions for the ordinary equation are correctly given, it can at worst be solved numerically, for which there are well-established techniques.

It was an almost fundamental assumption of classical physics that there would be a result to any conducted experiment, and this result should be reproducible within the limits of experimental error. This is the equivalent of assuming an existence theorem and a uniqueness theorem for the solution of the relevant partial differential equation. Hence the first concern of this book must be to examine the theorem and find out under what conditions it may be established. These considerations lead on to the more delicate questions as to what are the right forms of boundary conditions needed to solve particular equations. Intuitively one might suppose that the 'natural' boundary conditions for a problem in mathematical physics would also be the correct mathematical ones. Unfortunately this is not always the case. There are well-known branches of mathematical physics in which the complete formulation of the subject requires an equation, or system of equations, which must be accepted as empirically given, and a slight variation in these equations, physically acceptable, may change the character of the whole system. Thus, since the 'natural' boundary conditions will be unchanged, they may not be the correct ones for all the possible systems.

When we are satisfied that a given problem is correctly posed, that is that it has the correct mathematical boundary conditions, we may consider processes by which the equation may be solved. For some of the simplest equations of mathematical physics this may be possible analytically, and in later chapters we consider the form of such solutions. In general, however, it would probably be sufficient to obtain a numerical solution. We make no attempt to consider such work here, and refer the reader to textbooks of numerical analysis.

In this book we attempt no more than an introduction to the modern theory of partial differential equations. The reader wishing to find a more comprehensive statement of the modern theory should consult [1] or [4].

In the remainder of this chapter we look briefly at the equations which govern certain branches of mathematical physics. Our primary object is to see the contexts in which some of the equations arise, and so not too much attention has been paid to the underlying physics. For a more detailed treatment the reader is recommended to [23].

1.2 The equations of analytical dynamics

We consider first a system of particles, obeying Newton's laws of motion, and situated in a conservative field of force. Let there be n particles, of which the ith has position vector $\mathbf{x}_i = (x_i, y_i, z_i)$ and mass m_i. We assume further that there is a scalar potential function $V(t, \mathbf{x}_1, \mathbf{x}_2, \ldots, \mathbf{x}_n)$ and denote the vector with components

$$\left(\frac{\partial V}{\partial x_i}, \frac{\partial V}{\partial y_i}, \frac{\partial V}{\partial z_i}\right)$$

by $\text{grad}_i\, V$. Then the ordinary equations of motion are

$$\frac{d(m_i\dot{\mathbf{x}}_i)}{dt} + \text{grad}_i\, V = 0 \qquad (1.2.1)$$

where this repeated suffix is not summed.

If $\delta\mathbf{x}_i$ is an instantaneous displacement of the particle, consistent with the physical constraints, then from (1.2.1)

$$\sum_{i=1}^{n} (m_i\ddot{\mathbf{x}}_i . \delta\mathbf{x}_i + \text{grad}_i\, V . \delta\mathbf{x}_i) = 0. \qquad (1.2.2)$$

We introduce the kinetic energy

$$T = \tfrac{1}{2}\sum_{i=1}^{n} m_i(\dot{\mathbf{x}}_i)^2 \tag{1.2.3}$$

and assume that both this and V are expressed as functions of the time t and N generalized co-ordinates $q_1, q_2, \ldots, q_N$, and that T is also a quadric form in the generalized velocities $\dot{q}_1, \dot{q}_2, \ldots, \dot{q}_N$.

Now

$$\dot{\mathbf{x}}_i = \frac{\partial \mathbf{x}_i}{\partial t} + \sum_{j=1}^{N} \frac{\partial \mathbf{x}_i}{\partial q_j}\dot{q}_j$$

and so

$$\frac{\partial \dot{\mathbf{x}}_i}{\partial \dot{q}_k} = \frac{\partial \mathbf{x}_i}{\partial q_k}. \tag{1.2.4}$$

Hence

$$\frac{\partial T}{\partial \dot{q}_k} = \sum_{i=1}^{n} m_i\dot{\mathbf{x}}_i \cdot \frac{\partial \dot{\mathbf{x}}_i}{\partial \dot{q}_k} = \sum_{i=1}^{n} m_i\dot{\mathbf{x}}_i \cdot \frac{\partial \mathbf{x}_i}{\partial q_k} \tag{1.2.5}$$

and

$$\sum_{i=1}^{n} m_i\dot{\mathbf{x}}_i . \delta\mathbf{x}_i = \sum_{i=1}^{n}\sum_{k=1}^{N} m_i\dot{\mathbf{x}}_i . \frac{\partial \mathbf{x}_i}{\partial q_k}\delta q_k = \sum_{k=1}^{N}\frac{\partial T}{\partial \dot{q}_k}\delta q_k = \sum_{k=1}^{N} p_k\delta q_k \tag{1.2.6}$$

if as usual we define the generalized momentum by

$$p_k = \frac{\partial T}{\partial \dot{q}_k}. \tag{1.2.7}$$

When we differentiate (1.2.6) with respect to time we get

$$\frac{d}{dt}\left(\sum_{i=1}^{n} m_i\dot{\mathbf{x}}_i . \delta\mathbf{x}_i\right) = \sum_{i=1}^{n} m_i\dot{\mathbf{x}}_i . \delta\dot{\mathbf{x}}_i + \sum_{i=1}^{n} m_i\ddot{\mathbf{x}}_i . \delta\mathbf{x}_i = \frac{d}{dt}\left(\sum_{k=1}^{N} p_k\delta q_k\right)$$

or

$$\delta L = \delta T - \delta V = \frac{d}{dt}\left\{\sum_{k=1}^{N} p_k\,\delta q_k\right\} = \sum_{k=1}^{N}\{\dot{p}_k\,\delta q_k + p_k\,\delta\dot{q}_k\}, \tag{1.2.8}$$

from (1.2.2) and (1.2.3). Now $L = T - V$ is a function of t, q_k and $\dot{q}_k$, and its dependence on $\dot{q}_k$ is confined to the term T. Thus

$$\delta L = \sum_{k=1}^{N}\left\{\frac{\partial L}{\partial q_k}\delta q_k + \frac{\partial L}{\partial \dot{q}_k}\delta\dot{q}_k\right\} \tag{1.2.9}$$

and comparing (1.2.8) and (1.2.9) we have the Lagrange equations

$$\dot{p}_k = \frac{\partial L}{\partial q_k} \text{ and } p_k = \frac{\partial L}{\partial \dot{q}_k}. \tag{1.2.10}$$

We introduce the Hamiltonian function $H(t, q_k, p_k)$ which we assume dependent on t, the q_k and the generalized momenta p_k defined by

$$H(t, q_k, p_k) = \sum_{k=1}^{N} p_k \dot{q}_k - L. \tag{1.2.11}$$

Then

$$\begin{aligned}\delta H &= \sum_{k=1}^{N} \left\{\frac{\partial H}{\partial q_k}\delta q_k + \frac{\partial H}{\partial p_k}\delta p_k\right\},\\ &= \sum_{k=1}^{N} \{\dot{q}_k\,\delta p_k + p_k\,\delta q_k\} - \delta L = \sum_{k=1}^{N} \{\dot{q}_k\,\delta p_k - \dot{p}_k\,\delta q_k\},\end{aligned}$$

from (1.2.9) and (1.2.10). This gives Hamilton's canonical equations

$$\dot{q}_k = \frac{\partial H}{\partial p_k} \text{ and } \dot{p}_k = -\frac{\partial H}{\partial q_k}. \tag{1.2.12}$$

Finally, let

$$S = \int_{t_0}^{t} L\,dt. \tag{1.2.13}$$

We may suppose that it is possible to solve the ordinary differential equations for q_k, if we know the initial values q_{k_0}, $\dot{q}_{k_0}$. Thus for each k we may assume that q_k is a function of t, t_0, q_{k_0}, and $\dot{q}_{k_0}$. Similarly $\dot{q}_k$ can be expressed in terms of the same variables. These can be regarded as a system of algebraic equations which can be solved to express the variables $\dot{q}_k$ and $\dot{q}_{k_0}$ as functions of t, t_0, q_k and q_{k_0}. Then S can also be expressed in the same way, and so

$$\begin{aligned}\delta S &= \sum_{k=1}^{N} \left\{\frac{\partial S}{\partial q_{k_0}}\delta q_{k_0} + \frac{\partial S}{\partial q_k}\delta q_k\right\}\\ &= \int_{t_0}^{t} \delta L\,dt = \int_{t_0}^{t} \frac{d}{dt}\left\{\sum_{k=1}^{N} p_k\,\delta q_k\right\}dt = \sum_{k=1}^{N} \{p_k\,\delta q_k - p_{k_0}\,\delta q_{k_0}\},\end{aligned} \tag{1.2.14}$$

from (1.2.8). Furthermore

$$\begin{aligned}\frac{dS}{dt} &= \frac{\partial S}{\partial t} + \sum_{k=1}^{N} \frac{\partial S}{\partial q_k}\dot{q}_k,\\ &= L = \sum_{k=1}^{N} p_k \dot{q}_k - H.\end{aligned}$$

Therefore

$$\frac{\partial S}{\partial t} + H\left\{t, q_k, \frac{\partial S}{\partial q_k}\right\} = 0, \tag{1.2.15}$$

which is the Hamilton-Jacobi partial differential equation for S.

This equation is of first order and is one of the very few met in mathematical physics. We shall show, when we consider the first order equation, that a complete integral of (1.2.15) will involve $N + 1$ constants, which can be thought of as the initial time t_0 and the N values q_{k_0}. If another N constants β_k are introduced, Jacobi proved that first integrals of the equations of motion were given by

$$\frac{\partial S}{\partial q_{k_0}} = -\beta_k, \quad \frac{\partial S}{\partial q_k} = p_k.$$

A similar partial differential equation governs geometrical optics, in which the Lagrangian L is replaced by the refractive index of the medium in which the light is transmitted.

1.3 The equations of electromagnetism

We express the Faraday-Maxwell equations of the electromagnetic field in terms of Gaussian units. The fundamental physical quantities which specify the field are the scalar charge density ρ, and five vectors, $\mathbf{j}$ the current vector, $\mathbf{E}$ the electric field vector, $\mathbf{D}$ the electric displacement vector, $\mathbf{H}$ the magnetic field vector and $\mathbf{B}$ the magnetic induction vector.

Suppose we consider the electric field of a point charge *in vacuo*. Suppose further that a closed surface is drawn so that the point charge does not lie on the surface. We can consider the integral over this surface of the component of the electric field normal to the surface. It is shown in elementary treatises of potential theory that the integral is zero if the charge is outside the surface, or 4π if it is inside. This is an immediate consequence of the inverse-square law. It implies that

$$\operatorname{div} \mathbf{D} = 4\pi\rho. \tag{1.3.1}$$

Similarly it can be seen that

$$\operatorname{div} \mathbf{B} = 0. \tag{1.3.2}$$

(For the vector notation see [16].)

A change in the charge density implies that a current is flowing and so considering the balance of charge in a small volume we must have

$$\frac{\partial \rho}{\partial t} + \operatorname{div} \mathbf{j} = 0. \tag{1.3.3}$$

From (1.3.1) and (1.3.3)

$$\operatorname{div} \dot{\mathbf{D}} = 4\pi\dot{\rho} - 4\pi \operatorname{div} \mathbf{j},$$

and so

$$\dot{\mathbf{D}} + 4\pi\mathbf{j}$$

is solenoidal, where $\dot{\mathbf{D}}$ denotes $(\partial/\partial t)\mathbf{D}$.

Ampère's law says that the line integral of $\mathbf{H}$ round a closed curve is

$4\pi/c$ times the current enclosed, where c is the velocity of light, and this is equivalent to

$$c \operatorname{curl} \mathbf{H} = \dot{\mathbf{D}} + 4\pi\mathbf{j}, \tag{1.3.4}$$

since the right-hand side must be solenoidal.

Faraday's law of electromagnetic induction is that the line integral of **E** round a closed curve is equal to $-1/c$ times the rate of change of the flux of **B** through the curve, or

$$c \operatorname{curl} \mathbf{E} = -\dot{\mathbf{B}}. \tag{1.3.5}$$

These equations as they stand are insufficient to allow the field quantities to be found, and they must be supplemented by other relations whose validity is much more suspect than those above. Typical of these relations are

$$\mathbf{D} = \kappa\mathbf{E} \tag{1.3.6}$$

where κ is the dielectric constant,

$$\mathbf{B} = \mu\mathbf{H} \tag{1.3.7}$$

where μ is the magnetic permeability, and

$$\mathbf{j} = \sigma\{\mathbf{E} + \mathbf{E}_v\} \tag{1.3.8}$$

where σ is the specific conductivity, and $\sigma\mathbf{E}_v$ is the part of the current caused by the motion of the charges.

If μ is constant, the field is steady and currents are absent, (1.3.4) reduces to

$$\operatorname{curl} \mathbf{H} = 0, \tag{1.3.9}$$

which implies that there is a magnetic potential function ϕ_m so that

$$\mathbf{H} = -\operatorname{grad} \phi_m \tag{1.3.10}$$

and then (1.3.2) becomes

$$\nabla^2\phi_m = 0, \tag{1.3.11}$$

the Laplace equation. Similarly if the field is steady and κ is constant, the right-hand side of (1.3.5) vanishes and so

$$\mathbf{E} = -\operatorname{grad} \phi_e,$$

and from (1.3.1)

$$\kappa\nabla^2\phi_e = -4\pi\rho, \tag{1.3.12}$$

the Poisson equation.

Generally (1.3.2) implies that there is a field vector **A** so that

$$\mathbf{B} = \operatorname{curl} \mathbf{A} \tag{1.3.13}$$

and then from (1.3.5)

$$\operatorname{curl}\left\{c\mathbf{E} + \frac{\partial \mathbf{A}}{\partial t}\right\} = 0. \tag{1.3.14}$$

Hence there is a scalar function ϕ so that

$$c\mathbf{E} + \frac{\partial \mathbf{A}}{\partial t} = -c \operatorname{grad} \phi. \tag{1.3.15}$$

If we now assume that κ and μ are constant, from (1.3.4)

$$c^2 \operatorname{curl} \operatorname{curl} \mathbf{A} = \mu\kappa c\frac{\partial \mathbf{E}}{\partial t} + 4\pi c\mathbf{j} = 4\pi c\mathbf{j} - \mu\kappa\left\{\frac{\partial^2 \mathbf{A}}{\partial t^2} + c \operatorname{grad} \frac{\partial \phi}{\partial t}\right\}. \tag{1.3.16}$$

From (1.3.1) we obtain also

$$\kappa \operatorname{div}\left\{\frac{\partial \mathbf{A}}{\partial t} + c \operatorname{grad} \phi\right\} = -4\pi c\rho. \tag{1.3.17}$$

We may, without loss of generality, assume the Lorentz relation

$$\frac{\kappa\mu}{c}\frac{\partial \phi}{\partial t} + \operatorname{div} \mathbf{A} = 0, \tag{1.3.18}$$

for from (1.3.13) we may add to $\mathbf{A}$ the term grad χ where χ is an arbitrary function, and then from (1.3.15) we must correspondingly add $-(1/c)(\partial\chi/\partial t)$ to ϕ. When these changes are made on the left-hand side of (1.3.17) it is easily seen that χ may be so chosen to ensure that the right-hand side is zero.

Thus

$$\kappa\mu\frac{\partial^2 \phi}{\partial t^2} - c^2\nabla^2\phi = \frac{4\pi c^2}{\kappa}\rho \tag{1.3.19}$$

and

$$\kappa\mu\frac{\partial^2 \mathbf{A}}{\partial t^2} + c^2\{\operatorname{curl} \operatorname{curl} \mathbf{A} - \operatorname{grad} \operatorname{div} \mathbf{A}\} = 4\pi\mu c\mathbf{j}. \tag{1.3.20}$$

The term grad div − curl curl is a vector operator, which in the particular case of Cartesian coordinates reduces to the Laplacian ∇^2 acting on the Cartesian components of the vector. See in particular [16]. Hence (1.3.19) and (1.3.20) are equivalent to four inhomogeneous wave equations for ϕ and the three Cartesian components A_x, A_y, A_z.

Since light waves are electromagnetic, these wave equations may be considered as giving the equation which in wave optics corresponds to the equation of geometric optics. By analogy the corresponding equation to (1.2.15) is

$$\frac{\hbar}{i}\frac{\partial \psi}{\partial t} + H\left\{t, q_k, \frac{\hbar}{i}\frac{\partial}{\partial q_k}\right\}\psi = 0 \tag{1.3.21}$$

which is the Schrödinger equation of wave mechanics.

1.4 The equations of elasticity

We may derive the equations of motion of an elastic solid in the following manner. Let $\mathbf{x} = \{x_i\}$, $i = 1, 2, 3$, be the position vector of a point P of the body, and $\{x_i + \delta x_i\}$ that of an adjacent point Q. In some displacement of the body $P \rightarrow P'$ where P' has position vector $\{x_i + q_i\}$. Hence $Q \rightarrow Q'$ and the co-ordinates of Q' are

$$x_i + q_i + \delta x_i + \delta q_i.$$

The displacement of Q is therefore

$$\{q_i + \delta q_i\} = \left\{q_i + \frac{\partial q_i}{\partial x_j}\delta x_j\right\} \tag{1.4.1}$$

(where we now use the summation convention) and since this is a vector, as are also $\{q_i\}$ and $\{\delta x_i\}$, then $\{\partial q_i/\partial x_j\}$ is a second-rank Cartesian tensor.

We may write

$$\frac{\partial q_i}{\partial x_j} = \frac{1}{2}\left\{\frac{\partial q_i}{\partial x_j} - \frac{\partial q_j}{\partial x_i}\right\} + \frac{1}{2}\left\{\frac{\partial q_i}{\partial x_j} + \frac{\partial q_j}{\partial x_i}\right\}. \tag{1.4.2}$$

Hence the right-hand side of (1.4.1) can be divided into the sum of three terms

$$\mathbf{q} \tag{1.4.3}$$

which is a pure translation,

$$\frac{1}{2}\left\{\left(\frac{\partial q_i}{\partial x_j} - \frac{\partial q_j}{\partial x_i}\right)\delta x_j\right\} = -\tfrac{1}{2}\,\mathrm{curl}\,\mathbf{q} \times \delta\mathbf{x}, \tag{1.4.4}$$

which is a pure rotation, and

$$\frac{1}{2}\left\{\left(\frac{\partial q_i}{\partial x_j} + \frac{\partial q_j}{\partial x_i}\right)\delta x_j\right\} = \{e_{ij}\,\delta x_j\}, \tag{1.4.5}$$

which is a pure strain. The strain tensor e_{ij} is clearly a second-rank symmetrical tensor.

Now consider an arbitrary surface dividing the body into two pieces. The two sides will, by Newton's third law, exert on each other a system of equal and opposite forces. Each system can by the usual theorem be reduced to a resultant force and a couple. In particular, if we take a small element of surface dS, there will act on it a resultant force $\mathbf{F}$ and a couple $\mathbf{G}$. We postulate that as $dS \rightarrow 0$, $\mathbf{G}/dS \rightarrow 0$, while $\mathbf{F}/dS$ tends to a finite limit which we call the stress.

Let us suppose that we take as our surface a plane whose normal is in the direction of the unit vector $\mathbf{n}$, then we can represent the force on dS by the vector

$$\{p_{nj}\,dS\}. \tag{1.4.6}$$

Consider now a small tetrahedron taken from the body, with three sides dS_1, dS_2 and dS_3 parallel to the coordinate plane, and let the fourth side be dS. Then the equilibrium equations for the tetrahedron are

$$p_{nj}\,dS = p_{1j}\,dS_1 + p_{2j}\,dS_2 + p_{3j}\,dS_3 = p_{ij}\,dS_i, j = 1, 2 \text{ or } 3. \tag{1.4.7}$$

Now
$$dS_i = n_i\, dS,$$
and so
$$p_{nj} = p_{ij}n_i. \tag{1.4.8}$$

The normal component of stress across the plane in which dS lies we may call p_{nn}, and by straightforward resolution
$$p_{nn} = p_{nj}n_j = p_{ij}n_in_j. \tag{1.4.9}$$

This, however, is the transformation law for a second-rank tensor, and similar equations apply to the tangential stresses on dS. Thus p_{ij} is a second-rank tensor, the stress tensor. It must also be symmetric, since otherwise the moment of the non-normal stress components would produce a rotation in every small element of the body. Since both stress and strain are Cartesian tensors of second rank, by their usual properties at each point of the body there will be three mutually perpendicular axes of principal strain, and three of principal stress.

The fundamental assumption of Hooke's law is that there is a linear relation between stress and strain. This would, in general, imply the existence of a fourth-rank Cartesian tensor A_{ijkl} so that
$$p_{ij} = A_{ijkl}e_{kl}. \tag{1.4.10}$$

If, however, we assume that the body is isotropic, we can considerably reduce the number of coefficients from the 81 implied by (1.4.10). Take a, b and c to be specific values of the suffices, then we can reduce (1.4.10) to
$$p_{aa} = Ae_{aa} + A'\{e_{bb} + e_{cc}\} + Be_{bc} + B'\{e_{ab} + e_{ac}\}, \tag{1.4.11}$$
and
$$p_{bc} = Ce_{aa} + C'\{e_{bb} + e_{cc}\} + De_{bc} + D'\{e_{ab} + e_{ac}\}. \tag{1.4.12}$$
A rotation of the axes through half a circle about the third axis is represented by replacing p_{bc} by $-p_{bc}$ and p_{ac} by $-p_{ac}$ and leaving the other components invariant. This can be verified from the tensor transformation formula. Similar results apply to the e_{ij} components. Hence also
$$p_{aa} = Ae_{aa} + A'\{e_{bb} + e_{cc}\} - Be_{bc} + B'\{e_{ab} - e_{ac}\},$$
and so
$$B = B' = 0. \tag{1.4.13}$$
Also
$$-p_{bc} = Ce_{aa} + C'\{e_{bb} + e_{cc}\} - De_{bc} + D'\{e_{ab} - e_{ac}\},$$
and so
$$C = C' = D' = 0. \tag{1.4.14}$$
Thus
$$p_{aa} = Ae_{aa} + A'\{e_{bb} + e_{cc}\}, \tag{1.4.15}$$
and
$$p_{bc} = De_{bc}. \tag{1.4.16}$$

The last relation implies that the non-diagonal terms of the stress and strain matrices vanish together, and so the principal axes of stress and strain coincide. Let the principal stresses be P_1, P_2 and P_3 and the principal strains ε_1, ε_2 and ε_3, then from (1.4.15)

$$P_1 = A\varepsilon_1 + A'(\varepsilon_2 + \varepsilon_3) \tag{1.4.17}$$

with two similar relations. In the particular case of the stretched bar $P_1 = E\varepsilon$, $P_2 = P_3 = 0$ while $\varepsilon_1 = \varepsilon$, $\varepsilon_2 = \varepsilon_3 = \sigma E$, where E is Young's modulus and σ is Poisson's coefficient. We may therefore rewrite (1.4.17) as

$$E\varepsilon_1 = (1 + \sigma)P_1 - \sigma(P_1 + P_2 + P_3). \tag{1.4.18}$$

This is a tensor relation, however, and so the relation between stress and strain must be

$$Ee_{ij} = (1 + \sigma)p_{ij} - \sigma\delta_{ij}p_{kk} \tag{1.4.19}$$

where δ_{ij} is the Kronecker delta.

If we now consider the equations of motion of the body, applying Newton's laws to a small element acted upon by a force per unit mass $\mathbf{F}_i$ we get

$$\rho\frac{\partial^2 q_i}{\partial t^2} = \rho F_i + \frac{\partial p_{ji\cdot}}{\partial x_j} \tag{1.4.20}$$

Now from (1.4.19)

$$p_{ij} = 2\mu e_{ij} + \lambda\delta_{ij}e_{kk} \tag{1.4.21}$$

where

$$2(1 + \sigma)\mu = E \text{ and } (1 + \sigma)(1 - 2\sigma)\lambda = E\sigma.$$

On substituting from (1.4.21) and (1.4.5) into (1.4.20) we get

$$\rho\frac{\partial^2\mathbf{q}}{\partial t^2} = (\lambda + \mu) \text{ grad div } \mathbf{q} + \mu\nabla^2\mathbf{q} + \rho\mathbf{F}. \tag{1.4.22}$$

No ambiguity is possible in the interpretation of $\nabla^2\mathbf{q}$, since we have taken Cartesian components throughout. Taking the divergence of (1.4.22) we see that

$$\rho\frac{\partial^2}{\partial t^2}\{\text{div } \mathbf{q}\} - (\lambda + 2\mu)\nabla^2\{\text{div } \mathbf{q}\} = \text{div } \{\rho\mathbf{F}\}, \tag{1.4.23}$$

which is an inhomogeneous wave equation. Further taking the curl of equation (1.4.22) we see that

$$\rho\frac{\partial^2}{\partial t^2}\{\text{curl } \mathbf{q}\} - \mu\nabla^2\{\text{curl } \mathbf{q}\} = \text{curl } \{\rho\mathbf{F}\}, \tag{1.4.24}$$

which taken in Cartesian coordinates gives a set of three wave equations.

1.5 The equations of fluid dynamics

Finally we shall consider the equations for the laminar motion of a continuous fluid. If the density of fluid is ρ and its velocity is $\mathbf{v}$, then the change of density in a fluid element can only result from mass flow into or out of the element. Hence we shall have an equation of continuity

$$\frac{\partial \rho}{\partial t} + \operatorname{div}\{\rho \mathbf{v}\} = 0. \tag{1.5.1}$$

If the fluid is subjected to external forces **F** per unit mass, then applying Newton's second law to the momentum of the element resolved in the ith direction, and using (1.5.1) we find that

$$\rho\left\{\frac{\partial v_i}{\partial t} + v_j\frac{\partial v_i}{\partial x_j}\right\} = \frac{\partial p_{ji}}{\partial x_j} + \rho F_i \tag{1.5.2}$$

where, as in Section 1.4 above, p_{ij} are the components of the stress tensor.

The principal difference between fluid dynamics and elasticity theory is that for a viscous fluid we assume, not Hooke's law, but Newton's law that the stress is a linear function of the rate of strain tensor. This then replaces (1.4.21) by

$$p_{ij} = \mu e_{ij} + \left\{\frac{\lambda}{2}e_{kk} - p\right\}\delta_{ij} \tag{1.5.3}$$

where

$$e_{ij} = \frac{1}{2}\left\{\frac{\partial v_i}{\partial x_j} + \frac{\partial v_j}{\partial x_i}\right\}, \tag{1.5.4}$$

and p is the ordinary hydrostatic pressure. It is usually assumed that

$$\lambda + \tfrac{2}{3}\mu = 0, \tag{1.5.5}$$

although there is a considerable doubt as to the physical validity of this assumption. The coefficient μ is called the coefficient of dynamic viscosity. If the fluid is compressible, some form of energy equation will also be required, together with an equation of state. The energy equation (see for example [10]) can be taken as:

$$\rho\left\{\frac{\partial E}{\partial t} + v_j\frac{\partial E}{\partial x_j}\right\} = \Phi - p\frac{\partial v_j}{\partial x_j} + \frac{\partial}{\partial x_j}\left\{k\frac{\partial T}{\partial x_j}\right\} \tag{1.5.6}$$

where E is the internal energy, T is the temperature, k the thermal conductivity and Φ the dissipation given by

$$6\Phi = \mu\{6(e_{23}^2 + e_{31}^2 + e_{12}^2) + (e_{11} - e_{22})^2 + (e_{22} - e_{33})^2 + (e_{33} - e_{11})^2\}.$$

Alternatively, in terms of the entropy S, (1.5.6) may be written

$$\rho T\left\{\frac{\partial S}{\partial t} + v_j\frac{\partial S}{\partial x_j}\right\} = \Phi + \frac{\partial}{\partial x_j}\left\{k\frac{\partial T}{\partial x_j}\right\}. \tag{1.5.7}$$

If the fluid is incompressible and its motion is steady, the continuity equation reduces to

$$\operatorname{div} \mathbf{v} = 0. \tag{1.5.8}$$

If, furthermore, viscosity effects are neglected and the forces are conservative so that

$$F_i = -\frac{\partial V}{\partial x_i} \tag{1.5.9}$$

then (1.5.2) reduces to

$$\operatorname{grad}\left\{\tfrac{1}{2}\mathbf{v}^2 + \frac{p}{\rho} + V\right\} = \mathbf{v} \times \operatorname{curl} \mathbf{v} \tag{1.5.10}$$

so that $\frac{1}{2}\mathbf{v}^2 + p/\rho + V$ is constant along the streamlines. If it has a constant value at infinity then $\mathbf{v} \times \operatorname{curl} \mathbf{v}$ will vanish throughout the fluid, and we may assume that $\operatorname{curl} \mathbf{v} = 0$. There is then a velocity potential ϕ so that

$$\mathbf{v} = -\operatorname{grad} \phi \tag{1.5.11}$$

and ϕ will satisfy the Laplace equation

$$\nabla^2\phi = 0. \tag{1.5.12}$$

If the fluid is compressible, but we still neglect viscosity and heat conduction and assume a steady motion, we get a similar relation to (1.5.10) with the left-hand side replaced by $\operatorname{grad}\{\frac{1}{2}\mathbf{v}^2 + \int dp/\rho + V\}$. Since also (1.5.7) implies that S is constant on streamlines, if there are constant conditions at infinity the flow will be both irrotational ($\operatorname{curl} \mathbf{v} = 0$) and homentropic, S constant everywhere. We may then assume the adiabatic equation of state

$$p \propto \rho^\gamma, \qquad (1 < \gamma < 2), \tag{1.5.13}$$

and if we again use (1.5.11), equation (1.5.12) is replaced by

$$a^2\nabla^2\phi - v_i v_j \frac{\partial^2\phi}{\partial x_i\, \partial x_j} = 0, \tag{1.5.14}$$

where a has the dimensions of velocity, is called the velocity of sound for reasons that will become obvious below, and satisfies

$$a^2 = \frac{dp}{d\rho} = \gamma\frac{p}{\rho}. \tag{1.5.15}$$

If we assume that the flow is a small disturbance of a uniform flow with speed U_0 in the direction 1, and if the steady value a_0 of a is not nearly equal to U_0, we may put

$$\phi = -U_0 x_1 + \phi' \tag{1.5.16}$$

and then neglecting squares and products of derivatives of ϕ' (1.5.14) reduces to

$$(1 - M_0{}^2)\frac{\partial^2\phi'}{\partial x_1{}^2} + \frac{\partial^2\phi'}{\partial x_2{}^2} + \frac{\partial^2\phi'}{\partial x_3{}^2} = 0 \tag{1.5.17}$$

where $a_0^2M_0^2 = U_0^2$. This can be reduced to the Laplace equation by an affine transformation if $M_0 < 1$, and to the wave equation if $M_0 > 1$.

Now suppose we have a fluid of density ρ_0 at rest, and it then suffers a small disturbance. We may assume that the disturbance is irrotational and neglect viscosity and heat conduction. If the small velocity is $\mathbf{v}$ and the density changes to $\rho_0 + \rho'$ to first order, equation (1.5.1) becomes

$$\frac{\partial \rho'}{\partial t} + \rho_0 \operatorname{div} \mathbf{v} = 0. \tag{1.5.18}$$

Similarly, neglecting the external forces, (1.5.2) becomes

$$\frac{\partial \mathbf{v}}{\partial t} + \frac{1}{\rho_0} \operatorname{grad} p = 0. \tag{1.5.19}$$

To the same order we may replace grad p by

$$\operatorname{grad} p = \left\{\frac{\partial p}{\partial \rho}\right\}_0 \rho' = a_0^2 \rho' \tag{1.5.20}$$

where suffix 0 denotes the value when $\rho = \rho_0$, so that

$$\frac{\partial^2 \rho'}{\partial t^2} = a_0^2 \nabla^2 \rho'. \tag{1.5.21}$$

This is the wave equation, and it shows that a small disturbance (which describes a sound wave) propagates through the fluid at the speed of sound.

If we neglect entirely the motion of the fluid, and put into (1.5.6)

$$E = C_v T \tag{1.5.22}$$

where C_v is the specific heat at constant volume, (1.5.6) reduces to

$$\rho \frac{\partial}{\partial t}\{C_v T\} = \frac{\partial}{\partial x_j}\left\{k \frac{\partial T}{\partial x_j}\right\}, \tag{1.5.23}$$

which is the equation of heat conduction.

Consider next the motion of a fluid which is incompressible but viscous, then from (1.5.2) and (1.5.3) we can obtain

$$\frac{\partial \mathbf{v}}{\partial t} + \operatorname{grad}(\tfrac{1}{2}\mathbf{v}^2) - \mathbf{v} \times \operatorname{curl} \mathbf{v} + \frac{1}{\rho} \operatorname{grad} p + \nu \operatorname{curl} \operatorname{curl} \mathbf{v} = \mathbf{F}, \tag{1.5.24}$$

where $\nu = \mu/\rho$ is the kinematic viscosity.

If the forces are conservative, and we take the curl of this equation, and put

$$\boldsymbol{\omega} = \operatorname{curl} \mathbf{v} \tag{1.5.25}$$

where $\boldsymbol{\omega}$ is the vorticity, then

$$\frac{\partial \omega_i}{\partial t} + \omega_j \frac{\partial v_i}{\partial x_j} - v_j \frac{\partial \omega_i}{\partial x_j} = \nu \frac{\partial^2 \omega_i}{\partial x_j\, \partial x_j}. \tag{1.5.26}$$

If the motion is so slow that products of **v** and **ω** can be neglected, this reduces to the diffusion equation

$$\frac{\partial \omega_i}{\partial t} = \nu \nabla^2 \omega_i \tag{1.5.27}$$

showing that vorticity will gradually diffuse through a fluid when viscosity is present. The terms we have neglected in (1.5.26) represent a convection of vorticity.

We may in conclusion obtain the equations of magneto-hydrodynamics by linking the set of equations above with those of Section 1.3. The three links between the systems are established by substituting into (1.3.8)

$$\mathbf{E}_v = \frac{\rho_e}{\sigma}\mathbf{v} + \mu_e(\mathbf{v} \times \mathbf{H}) \tag{1.5.28}$$

where ρ_e is the charge density at a point, and μ_e is the magnetic permeability, putting in (1.5.2)

$$\mathbf{F} = \rho_e \mathbf{E} + \mu_e(\mathbf{j} \times \mathbf{H}) \tag{1.5.29}$$

and adding to the right-hand side of (1.5.6) and (1.5.7) a term representing electromagnetic energy.

Since, however, it is usually true that the fluid velocity is small compared to the velocity of light, some simplification is possible by neglecting terms of order $\mathbf{v}^2/c^2$. This replaces (1.5.29) by

$$\mathbf{F} = -\mu_e \mathbf{H} \times \operatorname{curl} \mathbf{H} \tag{1.5.30}$$

and replaces the Faraday-Maxwell equation by the single equation

$$\frac{\partial \mathbf{H}}{\partial t} = \operatorname{curl}\{v \times H\} - \operatorname{curl}\{\nu_H \operatorname{curl} \mathbf{H}\} \tag{1.5.31}$$

where $\nu_H = (\sigma\mu_e)^{-1}$ is the magnetic viscosity. It has also frequently been the practice to put $\nu_H = 0$, which is equivalent to assuming that the fluid has infinite conductivity.

CHAPTER 2

The Fundamental Existence and Uniqueness Theorems

2.1 Definition of the Cauchy problem

As we saw at the beginning of Chapter 1, we are normally concerned with trying to find, not solutions of partial differential equations in the abstract, but solutions satisfying certain given boundary data. We shall see in some later chapters that, for some of the well-known equations of mathematical physics, whole classes of elementary solutions are known, but despite this it may still be a matter of considerable mathematical difficulty to find a solution satisfying some boundary conditions. The important problem therefore is always the so-called Cauchy problem, to find a solution of the equation satisfying the given boundary conditions. If the boundary conditions given are sufficient to allow this, we may say that the Cauchy problem is well posed; if they are not, then the problem is not well posed.

In the case of some special equations, for example the Laplace equation, it is possible to prove that any solution which satisfies correctly-given boundary conditions exists, and this will also be unique. These proofs, which we shall meet in later chapters, depend essentially on the particular equation considered. It is clearly desirable, however, to have a more general existence and uniqueness theorem of the sort that is available for ordinary differential equations (see e.g. [11]). Such a theorem is that of Cauchy and Kowalewski.

For the sake of simplicity we shall confine our proof to equations of the second order (like most of those we shall examine later) and in two independent variables. The generalization to equations of higher orders and in more independent variables may be made reasonably easily, or reference may be made to other books, [1], [4] or [8].

The problem, then, is as follows. Given an equation of second order for a function $u(x, y)$, coupled with the knowledge of the values that

$u(x, y)$ takes on some curve $y = y(x)$, and given also the values taken on it by the derivative of u normal to the curve, to show that, under suitable conditions, a unique solution is defined in a neighbourhood of a point on the curve. This, then, is the Cauchy problem, although we do not yet know if it is well posed.

2.2 The formal solution

We start with a number of assumptions.

(*a*) The curve is the axis of y, or $x = 0$.

(*b*) The equation may be solved for $\partial^2 u/\partial x^2$, so that it takes the simplified form

$$\frac{\partial^2 u}{\partial x^2} = f\left\{x, y, u, \frac{\partial u}{\partial x}, \frac{\partial u}{\partial y}, \frac{\partial^2 u}{\partial x\,\partial y}, \frac{\partial^2 u}{\partial y^2}\right\} \tag{2.2.1}$$

or

$$r = f(x, y, u, p, q, s, t). \tag{2.2.2}$$

(*c*) The function $f\{x, y, u, p, q, s, t\}$, regarded as a function of seven variables, is analytic, that is, it possesses partial derivatives of all orders, in some region which includes a segment of $x = 0$.

(*d*) If the boundary conditions are written in the form

$$u(x, y) = u_0(y) \text{ and } \frac{\partial u}{\partial x} = p = u_1(y) \text{ on } x = 0,$$

then both $u_0(y)$ and $u_1(y)$ are analytic functions of y in some range $|y - y_0| < h$, so that the rectangular region $|y - y_0| < h, x \geqslant 0$ overlaps with the region in the previous assumption.

We look for a solution of the form

$$u(x, y) = u_0(y) + xu_1(y) + \frac{x^2}{2!}u_2(y) + \frac{x^3}{3!}u_3(y) + \ldots \tag{2.2.3}$$

which clearly satisfies the boundary conditions, in the neighbourhood of $(0, y_0)$.

On the line $x = 0$,

$$q = u_0'(y),\ r = u_2(y),\ s = u_1'(y),\ t = u_0''(y),$$

where the primes denote differentiation with respect to y. If then the equation is satisfied we must have

$$u_2 = f(0, y, u_0, u_1, u_0', u_1', u_0''), \tag{2.2.4}$$

and since the arguments of f only involve u_0, u_1 and their derivatives, which are known, u_2 can be regarded as a known function of y.

Furthermore on $x = 0$, of the third derivatives

$$\frac{\partial^3 u}{\partial x^2\,\partial y} = \frac{\partial r}{\partial y} = u_2'(y), \frac{\partial^3 u}{\partial x\,\partial y^2} = \frac{\partial s}{\partial y} = u_1''(y), \frac{\partial^3 u}{\partial y^3} = \frac{\partial t}{\partial y} = u_0'''(y),$$

all are now known. Since, if we differentiate (2.2.2),

$$\frac{\partial^3 u}{\partial x^3} = \frac{\partial f}{\partial x} + p\frac{\partial f}{\partial u} + r\frac{\partial f}{\partial p} + s\frac{\partial f}{\partial q} + \frac{\partial r}{\partial y}\frac{\partial f}{\partial s} + \frac{\partial s}{\partial y}\frac{\partial f}{\partial t}, \tag{2.2.5}$$

$u_3(y)$ may be found. We can therefore successively calculate all the coefficients in the expansion (2.2.3), and so formally find a solution of our problem. Provided always that f is a single valued function of its arguments, at each stage the new coefficient is defined without ambiguity, and so the solution must be unique. The difficulty is to show that an expansion like (2.2.3) can converge in some region near the boundary. If it does, however, it will converge to a unique solution of our boundary value problem.

2.3 The dominant solution

The Cauchy-Kowalewski method is to construct a similar series, every term of which may be shown to dominate the equivalent term in our series, and whose convergence may be more easily established.

We suppose that $(0, y_0)$ is a point on the boundary in the neighbourhood of which f is analytic. It is convenient to work with a new function $u_*(x, y)$ rather than with $u(x, y)$. This function is defined by:

$$u_*(x, y) = u(x, y) - u_0(y) - xu_1(y) - \frac{x^2}{2}u_2(y), \tag{2.3.1}$$

so that, with corresponding definitions for the starred quantities, the equation becomes

$$r_* = f_*(x, y, u_*, p_*, q_*, s_*, t_*), \tag{2.3.2}$$

and on $x = 0$

$$0 = u_* = p_* = q_* = s_* = t_* = r_*. \tag{2.3.3}$$

The assumption of the analytic character of f and therefore of f_* is equivalent to assuming for it an infinite multi-Taylor series in powers of $x, y - y_0, u_*, p_*, q_*, s_*$ and t_*. We know that the moduli of the coefficients of a multi-Taylor series are bounded above by a result analogous to that of the Cauchy inequalities for the ordinary Taylor series. Thus $f_*(x, y, u_*, p_*, q_*, s_*, t_*)$ is dominated by an expression of the form

$$K\left[\left\{1 - \frac{x + y - y_0 + u_* + p_* + q_*}{\rho'}\right\}^{-1}\left\{1 - \frac{s_* + t_*}{\rho_1}\right\}^{-1} - 1\right] \tag{2.3.4}$$

for some constants K, ρ' and ρ_1, in the sense that in its respective expansion each coefficient in the latter expansion (2.3.4) dominates its corresponding coefficient. We may furthermore use a trick first employed by Goursat and notice that the dominant character of (2.3.4) is further enhanced if x is replaced by x/α, where α is positive and less than 1. We are led to examine the solution of an equation

$$R_* = K\left[\left\{1 - \frac{x/\alpha + y - y_0 + U_* + P_* + Q_*}{\rho'}\right\}^{-1} \times \left\{1 - \frac{S_* + T_*}{\rho_1}\right\}^{-1} - 1\right]$$

$$= F_*(x, y, U_*, P_*, Q_*, S_*, T_*) \qquad (2.3.5)$$

for a function $U_*(x, y)$ where by analogy we have put $P_* = \partial U_*/\partial x$, $Q_* = \partial U_*/\partial y$ etc. There are solutions of (2.3.5) which are functions only of the single independent variable

$$\xi = x + \alpha(y - y_0).$$

Denoting by primes differentiation with respect to ξ, (2.3.5) becomes

$$U_*'' = K\left[\left\{1 - \frac{\xi/\alpha + U_* + (1 + \alpha)U_*'}{\rho'}\right\}^{-1} \times \left\{1 - \left(\frac{\alpha + \alpha^2}{\rho_1}\right)U_*''\right\}^{-1} - 1\right], \qquad (2.3.6)$$

or

$$\left[1 - \frac{K(\alpha + \alpha^2)}{\rho_1}\right]U_*'' - \frac{\alpha + \alpha^2}{\rho_1}\{U_*''\}^2 = K\left[\left\{1 - \frac{\xi/\alpha + U_* + (1 + \alpha)U_*'}{\rho'}\right\}^{-1} - 1\right]. \qquad (2.3.7)$$

If we assume that at $\xi = 0$, $U_* \geqslant 0$, $U_*' \geqslant 0$, U'' will also be $\geqslant 0$ if we select α so that $1 - K(\alpha + \alpha^2)/\rho_1 > 0$, which we can always do. Equation (2.3.5) is similar in its form to (2.2.2). Hence by a process of differentiation like that exhibited in (2.2.5) we could construct a formal series solution for U_*, similar to (2.2.3). The coefficients would be generated successively, just as were those of (2.2.3), by successive differentiations, expressed by formulae like (2.2.4) with partial derivatives of F_* replacing those of f_*, and partial derivatives of U_* replacing those of u_*. By the construction of our dominant solution all the partial derivatives of F_* are positive and are greater than or equal to the moduli of the corresponding partial derivatives of f_*. We know also that at $(0, y_0)$, $u_* = p_* = q_* = s_* = t_* = r_* = 0$, while $U_* = 0$, and $P_* \geqslant 0$, $Q_* \geqslant 0$, $R_* \geqslant 0$, $S_* \geqslant 0$ and $T_* \geqslant 0$. It follows therefore that $U_{*3}(y) \geqslant |u_3(y)|$ and, by employing similar arguments, successively that $U_{*n}(y) \geqslant |u_n(y)|$. The coefficients of the series for $U_*(x, y)$ are therefore greater than those in the series for $u(x, y)$ and so the region in which the series for $u(x, y)$ converges is at least as large as that in which the series for $U_*(x, y)$ converges. We have, however, reduced the problem of finding $U_*(x, y)$ to the solution of an ordinary differential equation of the second order, and for these we already have existence theorems [11]. Then the solution for $U_*(x, y)$ exists in some neighbourhood of $(0, y_0)$ and so a solution for $u(x, y)$ must exist in at least as large a neighbourhood. This completes the proof.

2.4 Examination of the hypotheses

Since our proof was constructed on the basis of the four assumptions (*a*), (*b*), (*c*) and (*d*), we must now return and examine them in greater detail.

In general the boundary conditions will be given on a curve $y = y(x)$, and not on $x = 0$. If, however, this curve has a continuously turning tangent, at every point of some segment we can find a family of curves, of which our boundary curve is a member, so that a unique member of the family passes through every point of some region including the segment of the curve. We can then take this family of curves and their orthogonal trajectories as giving a new orthogonal coordinate grid covering our region. A change of independent variables to the new ones so defined will then reduce the general boundary condition to that assumed in (*a*).

In place of assumption (*b*) we would have to assume that the equation was given as

$$\phi(x, y, u, p, q, r, s, t) = 0 \qquad (2.4.1)$$

where ϕ is a function of all eight variables, which, if we still maintain a hypothesis similar to (*c*), possesses partial derivatives of all orders in some region. Then by the implicit function theorem [6], the condition that in the neighbourhood of some point of the eight-dimensional space, ϕ can be solved for r to give an equation like (2.2.1) is just:

$$\frac{\partial\phi}{\partial r} \neq 0$$

at the point. This may be interpreted intuitively as a condition that we are able to integrate step by step away from the boundary. We have seen that the boundary conditions that both u and p are given on $x = 0$ also imply the knowledge of q, s and t on the line, so that, in general,

$$\frac{\partial\phi}{\partial r} = 0$$

at some point will imply a relationship between these known values. In the particular case of a 'semi-linear' second-order equation

$$A(x, y)r + 2B(x, y)s + C(x, y)t = D(x, y, u, p, q) \qquad (2.4.2)$$

the condition is simply

$$A(0, y) \neq 0.$$

If, however, the boundary condition for our semi-linear equation is given on a curve $y = y(x)$, then, as we have said above, we may introduce new orthogonal curvilinear coordinates

$$\xi = \xi(x, y), \quad \eta = \eta(x, y),$$

so that $y = y(x)$ is given by $\xi = 0$. After we have transformed our equation, the corresponding condition will be that the coefficient of $\partial^2 u/\partial \xi^2$ must not vanish on $\xi = 0$, which can easily be seen as:

$$A(x, y)\left(\frac{\partial \xi}{\partial x}\right)^2 + 2B(x, y)\frac{\partial \xi}{\partial x}\frac{\partial \xi}{\partial y} + C\left(\frac{\partial \xi}{\partial y}\right)^2 \neq 0. \qquad (2.4.3)$$

If its left-hand side is zero, (2.4.3) may be interpreted as a differential equation to determine the function $\xi(x, y)$; if, in the (x, y) plane, there exists a family of real curves obtained by putting $(x, y) =$ const. we shall call these the characteristic curves of (2.4.2). Clearly, from what we have said, they will represent those curves from which we would be unable to integrate the equation with the given boundary conditions only. Furthermore we have also seen that on such curves some kind of relationship must be satisfied by these boundary data.

2.5 Applicability of the theorem

We come finally to the assumptions (*c*) and (*d*), and it is fairly obvious that, in the type of existence proof that has been used, both are essential. Neither could be dropped without the whole form of the proof having to be recast.

On purely mathematical grounds it would be desirable to be able to relax assumption (*c*), but it must be freely admitted that the equations that the mathematical physicist is required to consider usually are analytic except, at most, in regions of measure zero.

Assumption (*d*) is, however, of a much more restrictive character. We have taken the boundary conditions to be that both u and its normal derivative have been given on a plane or surface, which conditions we can summarize as being given two bits of information at every point of an open boundary. We can easily construct simple physical problems, however, in which the boundary conditions are not of this type. An example would be the determination of the electrostatic potential in a region of space bounded by conductors when the potentials of the conductors were given, or, if the region extended to infinity, when the behaviour of the potential at large distances was given. In fact, as we shall see in a later chapter, any problem requiring the solution of the Laplace equation will have just one piece of information given at every point of a closed boundary surrounding the region concerned.

Even, however, for physical problems whose natural boundary conditions are of the type assumed in this chapter, the restriction that both u and its normal derivative shall themselves be analytic is far too great. After all, any problem wherein some new process is started suddenly, at a finite instant of time, will involve some kind of discontinuity, and cannot therefore be analytic.

We shall see in Chapter 4 that what is required is a deeper insight into the behaviour of partial differential equations than is given by this existence and uniqueness theorem.

Exercises

2.1 Construct formally a solution of

$$\frac{\partial^2 u}{\partial x^2} = \frac{\partial^2 u}{\partial y^2}$$

as a series

$$u(x, y) = u_0(y) + u_1(y)x + \frac{u_2(y)}{2}x^2 + \ldots + \frac{u_n(y)}{n!}x^n + \ldots,$$

given that on $x = 0$, $u_0(y) = \exp(-y^2)$, $u_1(y) = -2y \exp(-y^2)$.

Show that this is, for any fixed y, a Taylor series for $\exp\{-(x + y)^2\}$, and hence can be continued over the whole plane.

2.2. Construct a formal solution in series of the above type for the equation

$$y^2\frac{\partial^2 u}{\partial x^2} = x^2\frac{\partial^2 u}{\partial y^2} + 2(x^2 - y^2)u$$

given that $u_0(y) = \exp(-y^2)$, $u_1(y) = 0$. Hence show that for every n, $u_{2n+1}(y) = 0$ while $u_{2n}(y)$ is of the form $c_n u_0(y)$, for some constant c_n, and deduce that

$$u(x, y) = \exp\{-(x^2 + y^2)\}.$$

CHAPTER 3

First-order Equations

3.1 Definitions and notation

We shall consider equations in the independent variables $x_1, x_2, \ldots, x_n$, which we shall, when convenient, denote collectively by a vector $\mathbf{x}$. The equation will determine a function $u(\mathbf{x})$, and we put $p_i = \partial u/\partial x_i$, and will denote by $\mathbf{p}$ the whole set of first partial derivatives. It will sometimes be convenient, however, to consider two-dimensional problems separately, and then the independent variables will be taken as x and y and in the usual convention $p = \partial u/\partial x$, $q = \partial u/\partial y$.

A first-order equation will be of the form

$$\phi(\mathbf{x}, u, \mathbf{p}) = 0. \tag{3.1.1}$$

We shall say that the equation is linear if it has the simpler form:

$$\sum_{j=1}^{n} A_j(\mathbf{x})p_j + B(\mathbf{x})u = C(\mathbf{x}); \tag{3.1.2}$$

that it is semi-linear if of the form

$$\sum_{j=1}^{n} A_j(\mathbf{x})p_j = C(\mathbf{x}, u), \tag{3.1.3}$$

and that it is quasi-linear if

$$\sum_{j=1}^{n} A_j(\mathbf{x}, u)p_j = C(\mathbf{x}, u). \tag{3.1.4}$$

3.2 The linear equation in two independent variables

Consider first the simple linear equation, in two independent variables

$$A_1(x, y)p + A_2(x, y)q = 0. \tag{3.2.1}$$

We may compare this with the differential relation

$$p\,dx + q\,dy = du. \tag{3.2.2}$$

If the incremental changes dx and dy are chosen in such a direction that

$$\frac{dx}{A_1(x, y)} = \frac{dy}{A_2(x, y)} \tag{3.2.3}$$

then (3.2.2) tells us that the corresponding incremental change in u vanishes. Now (3.2.3) can be identified with the ordinary first-order differential equation

$$\frac{dy}{dx} = \frac{A_2(x, y)}{A_1(x, y)}. \tag{3.2.4}$$

Subject, therefore, to the restrictions placed on $A_1(x, y)$ and $A_2(x, y)$ by the requirements of the existence theorem for ordinary first-order equations [11], there will be a unique integral curve of (3.2.4) passing

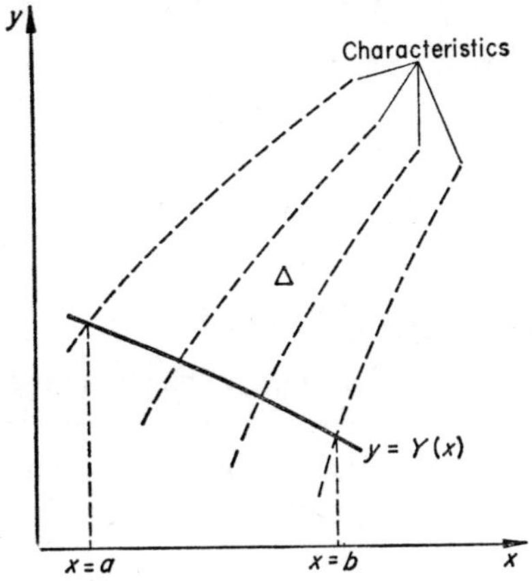

Fig. 3.1

through any given point (x_0, y_0). The exceptions will be the set of singular points of the equation satisfying

$$A_1(x, y) = A_2(x, y) = 0. \tag{3.2.5}$$

We shall call an integral curve of (3.2.4) a characteristic of (3.2.1).

The Cauchy problem will be to find a solution of (3.2.1) which satisfies a boundary condition that $u(x, y)$ is known on some segment of a given curve, say

$$u(x, y) = U(x) \quad \text{on} \quad y = Y(x) \quad \text{for} \quad a \leqslant x \leqslant b. \tag{3.2.6}$$

We suppose that no point of the curve segment satisfies simultaneously the pair of equations (3.2.5), or, in other words, no singular point of the equation lies on the arc. Then either through each point of the arc passes a unique integral curve of (3.2.4) and the whole set covers some region Δ of the (x, y) plane (see Fig. 3.1), or the initial curve is itself a characteristic.

If the initial curve $y = Y(x)$ is itself a characteristic, then since we know that u is constant on the characteristics, we must have that $U(x)$ is constant, for consistency. We cannot integrate away from the initial line, since it follows from (3.2.3) that the differentiation implied by (3.2.2) is along this initial line. When $U(x)$ is constant, so that the boundary data are consistent, we can construct a solution of (3.2.1) with u constant on each characteristic, where the constant values are arbitrary except in the case of the initial line. Thus there is no unique solution of the Cauchy problem. If $U(x)$ is not constant, the boundary data are inconsistent and so no solution whatsoever is possible.

The integral curves of (3.2.4) will be a one-parameter family of the form

$$y = y(x, C). \tag{3.2.7}$$

If then (x_0, y_0) is a point of Δ, there is a unique value of C, say C_0, so that

$$y_0 = y(x_0, C_0). \tag{3.2.8}$$

The particular integral curve $y = y(x, C_0)$ will meet the initial curve $y = Y(x)$ in a unique point (x', y') where

$$Y(x') = y(x', C_0) \quad \text{and} \quad y' = Y(x'). \tag{3.2.9}$$

We have seen however that $u(x, y)$ is constant along the integral curves, and so the value that $u(x, y)$ has at (x_0, y_0) is the same as that at (x', y'). This value is $U(x')$, and is known. Thus we see that the value of $u(x, y)$ is determined uniquely by the boundary condition at each point of Δ.

3.3 The semi-linear equation in two independent variables

Little substantial change is introduced if we replace (3.2.1) by the semi-linear equation

$$A_1(x, y)\frac{\partial u}{\partial x} + A_2(x, y)\frac{\partial u}{\partial y} = C(x, y, u). \tag{3.3.1}$$

As before, we have a one-parameter family of characteristics, defined independently of the boundary conditions by the equation (3.2.4), whose equations we can take as being given by (3.2.7), and so, provided the initial line is not characteristic, the Cauchy problem can be solved uniquely.

In this case, however, comparison of (3.3.1) with (3.2.2) leads to the conclusion that, on a characteristic of the family,

$$\frac{du}{C(x, y, u)} = \frac{dx}{A_1(x, y)} = \frac{dy}{A_2(x, y)}. \tag{3.3.2}$$

If we associate parameter C_0 with a particular characteristic, we can therefore see from (3.3.2) that on that characteristic

$$\frac{du}{dx} = \frac{C\{x, y, u\}}{A_1\{x, y\}} = \frac{C\{x, y(x, C_0), u\}}{A_1\{x, y(x, C_0)\}}, \tag{3.3.3}$$

which is an ordinary first-order equation to find u. To solve our Cauchy problem then, we again extrapolate the characteristic back from our given point (x_0, y_0) until it meets the initial curve $y = Y(x)$ at (x', y'). Then we integrate (3.3.3) forward, taking the boundary condition that u has the value $U(x')$ at $x = x'$, and this will in general determine a unique value of u at $x = x_0$, which is the solution of the Cauchy problem.

3.4 The quasi-linear equation in two independent variables

We could proceed in a somewhat similar manner in the case of the quasi-linear equation

$$A_1(x, y, u)\frac{\partial u}{\partial x} + A_2(x, y, u)\frac{\partial u}{\partial y} = C(x, y, u). \tag{3.4.1}$$

Now, however, the equation to determine the characteristics involves u, and so we must solve the equations equivalent to (3.2.4) and (3.3.3), namely

$$A_1\{x, y, u\}\frac{dy}{dx} = A_2\{x, y, u\} \text{ and } A_1\{x, y, u\}\frac{du}{dx} = C\{x, y, u\}, \tag{3.4.2}$$

together as a pair of simultaneous equations. Clearly in this case the shape of the characteristic will depend on the given boundary conditions, since they are influenced by the value of u, and will not be determined by the partial differential equation alone as in the cases previously considered. It follows therefore that the region Δ, that is the region in which a solution can be determined by the boundary conditions, must now depend on those boundary conditions.

We can, however, regard (3.4.1) alternatively as being a linear equation, of the simple type of (3.2.1), but in a three-dimensional space with coordinates (x, y, u). If we assume that u is given as a function of x and y by the implicit relation

$$f(x, y, u) = \text{const.} \tag{3.4.3}$$

then the differential relation is

$$df = \frac{\partial f}{\partial x}dx + \frac{\partial f}{\partial y}dy + \frac{\partial f}{\partial u}du = 0. \tag{3.4.4}$$

Comparing this with (3.2.2) we see that

$$\frac{\partial f/\partial u}{-1} = \frac{\partial f/\partial x}{p} = \frac{\partial f/\partial y}{q} \tag{3.4.5}$$

and when this is substituted in (3.4.1), it reduces to the simple linear equation

$$A_1\{x, y, u\}\frac{\partial f}{\partial x} + A_2\{x, y, u\}\frac{\partial f}{\partial y} + C\{x, y, u\}\frac{\partial f}{\partial u} = 0. \qquad (3.4.6)$$

This can always be generalized, so that a quasi-linear equation in n dimensions may be regarded as a homogeneous linear equation in an $(n + 1)$-dimensional space.

3.5 The linear equation in *n* independent variables

The most general Cauchy problem for a quasi-linear equation of first order can therefore be regarded as equivalent to finding those characteristics of a homogeneous linear equation in one or more independent variables, which cut a given twisted curve. Hence we consider the equation

$$\sum_{j=1}^{n} A_j(\mathbf{x})p_j = 0. \qquad (3.5.1)$$

If we compare this with the differential relation

$$du = \sum_{j=1}^{n} p_j dx_j, \qquad (3.5.2)$$

we see that the characteristics are lines on which u is constant. These satisfy the system of ordinary differential equations

$$\frac{dx_j}{A_j(\mathbf{x})} = \frac{ds}{A(\mathbf{x})} \quad j = 1, 2, \ldots, n \qquad (3.5.3)$$

where
$$\{A(\mathbf{x})\}^2 = \sum_{j=1}^{n} \{A_j(\mathbf{x})\}^2.$$

We may assume that the given curve is specified in terms of a single parameter σ, so that $x_j = x_j(\sigma)$ describes this curve.

Let $\mathbf{x}' = \mathbf{x}(\sigma')$ be a point on the initial twisted curve corresponding to $\sigma = \sigma'$. We can under suitable conditions find solutions of the equation set (3.5.3) so that $\mathbf{x} = \mathbf{x}'$ when $s = 0$. The solution will then be of the form

$$\mathbf{x} = \mathbf{x}(s, \sigma'). \qquad (3.5.4)$$

which as s varies will be the coordinate vector of a point which moves on a twisted curve containing the point $\mathbf{x}'$. Once again therefore we find that either the integral curve coincides with the initial curve, or it meets it in one point. If the initial curve is an integral curve of the system of equations (3.5.3) then the Cauchy problem has no solution. If not, then the set of characteristics which intersects any segment of our initial curve will generate a piece of surface

$$\mathbf{x} = \mathbf{x}(s, \sigma) \quad a \leqslant \sigma \leqslant b, \tag{3.5.5}$$

and this surface represents the solution of the Cauchy problem.

3.6 General first-order equation in two independent variables

In this section we shall take the dependent variable as z, instead of u as in the earlier parts of the chapter, so that any value of a solution for particular values of x and y will be represented by a point in three-dimensional space. We shall still write $p = \partial z/\partial x$ and $q = \partial z/\partial q$ so that a general first-order equation will have the form

$$F(x, y, z, p, q) = 0. \tag{3.6.1}$$

A solution of (3.6.1) will be some relation between z and the independent variables, and so will be represented by a surface in space. The Cauchy problem will take the form of finding such a surface which contains a given twisted curve in space.

Any function $z = z(x, y, a, b)$ which satisfies (3.6.1) and contains two arbitrary constants is a 'complete integral' of the equation. If the surfaces represented by the complete integral pass through a given point (x_0, y_0, z_0) this will imply a functional dependence of the constants a and b.

At any point on an integral surface the direction cosines of the normal to the surface are in the ratio $p : q : -1$ and so (3.6.1) represents a relation between the direction cosines to the integral surface at every given point. Consider a given point P with coordinates x, y, z, and an integral surface S passing through it. The tangent plane to S at P has the equation

$$\zeta - z = p(\xi - x) + q(\eta - y) \tag{3.6.2}$$

where (ξ, η, ζ) are running coordinates in the plane.

Since, however, (3.6.1) gives a single relation between p and q at the point P, the tangent planes to all integral surfaces through P form a one-parameter family and will envelop a cone, with vertex at P. Let us call this the tangent cone $T(P)$, since it clearly depends only on (3.6.1) and the particular point P. A tangent plane (3.6.2) to S is also a tangent plane to $T(P)$.

The cone $T(P)$ is generated by the intersection of neighbouring tangent planes, and so we must consider the intersection of (3.6.2) with

$$\zeta - z = (p + \delta p)(\xi - x) + (q + \delta q)(\eta - y) \tag{3.6.3}$$

or if we subtract (3.6.2) and pass to the limit we get

$$\xi - x + (\eta - y)\frac{dq}{dp} = 0. \tag{3.6.4}$$

But from (3.6.1)

$$\frac{\partial F}{\partial p} + \frac{\partial F}{\partial q}\frac{dq}{dp} = 0,$$

and so eliminating dq/dp

$$(\eta - y)\frac{\partial F}{\partial p} - (\xi - x)\frac{\partial F}{\partial q} = 0. \tag{3.6.5}$$

The equation of the cone $T(P)$ is then formally obtained by eliminating p and q between the equations (3.6.1), (3.6.2) and (3.6.5). Since in fact (3.6.1) is independent of (ξ, η, ζ) and in the other equations it enters merely in the combinations $\xi - x$, $\eta - y$, $\zeta - z$, it follows that if some point lies on $T(P)$ so does every point on the line joining it to the vertex P, which is therefore a generator of the cone.

The normal to S at P has the equation

$$\frac{\xi - x}{p} = \frac{\eta - y}{q} = \frac{\zeta - z}{-1} \tag{3.6.6}$$

and so the set of all normals generates another cone, the normal cone $N(P)$ whose equation is

$$F\left(x, y, z, \frac{x - \xi}{\zeta - z}, \frac{y - \eta}{\zeta - z}\right) = 0. \tag{3.6.7}$$

In the cases we have treated previously of the linear, semi-linear or quasi-linear equation, F was linear in p and q, and so the cone $T(P)$ was a line, and $N(P)$ was a plane. The line into which $T(P)$ degenerates would be a tangent at each point to the characteristic through the point.

In the general case, we have seen that at each point P of an integral surface the tangent plane touches the cone $T(P)$ along a generator of the cone. Except possibly for singular points, this generator is uniquely defined, and so we can define the characteristic curves on the integral surfaces as curves which everywhere have this generator as a tangent. This is obviously consistent with the previous sections. In fact, associated with each point of the characteristic curve there will not only be a tangent generator, but also a particular tangent plane of $T(P)$, and so there is also defined a strip associated with the characteristic. A set of values x, y, z, p, q satisfying (3.6.1) is said to be an element of the strip.

Let ξ, η, ζ be a point on the generator of the cone at P. From (3.6.2) and (3.6.5)

$$\frac{\xi - x}{P} = \frac{\eta - y}{Q} = \frac{\zeta - z}{pP + qQ},$$

where in the usual notation we have put $P = \partial F/\partial p$ and $Q = \partial F/\partial q$. These generators are tangent to the characteristic curves, and so if we let $\xi \to x$, $\eta \to y$, $\zeta \to z$ we derive as equations of the characteristic curve:

$$\frac{dx}{P} = \frac{dy}{Q} = \frac{dz}{pP + qQ}. \tag{3.6.8}$$

If in the usual notation we write the second derivatives

$$r = \frac{\partial^2 z}{\partial x^2}, \quad s = \frac{\partial^2 z}{\partial x\, \partial y}, \quad t = \frac{\partial^2 z}{\partial y^2},$$

then clearly we have differential relations

$$dp = r\,dx + s\,dy \tag{3.6.9}$$

and

$$dq = s\,dx + t\,dy. \tag{3.6.10}$$

If, further, we differentiate (3.6.1) with respect to x we get

$$X + pZ + rP + sQ = 0, \tag{3.6.11}$$

and with respect to y

$$Y + qZ + sP + tQ = 0, \tag{3.6.12}$$

where we have put $X = \partial F/\partial x$, $Y = \partial F/\partial y$, $Z = \partial F/\partial z$.

From (3.6.8), (3.6.9) and (3.6.10)

$$P\,dp = (rP + sQ)\,dx$$

and

$$Q\,dq = (sP + tQ)\,dy,$$

and so with the aid of (3.6.11) and (3.6.12) we can extend the equations (3.6.8) to obtain Charpit's equation for the characteristic strip

$$\frac{dx}{P} = \frac{dy}{Q} = \frac{dz}{pP + qQ} = -\frac{dp}{X + pZ} = -\frac{dq}{Y + qZ}. \tag{3.6.13}$$

The Charpit relations give five first-order ordinary differential equations for x, y, z, p and q in terms of some single parameter. In general therefore the strip is uniquely defined if one set of values x_0, y_0, z_0, p_0, q_0 is given. Hence if two integral surfaces touch at one point, they touch at every point of a characteristic curve.

The set of equations (3.6.13) has, in general, a triple infinity of solutions. This can be seen from the fact that any initial value set x_0, y_0, z_0, p_0, q_0 must satisfy (3.6.1), and also if we regard them as the equations of trajectories in a five-dimensional space we can define them by their intersection with some particular plane, say $x = x_0$.

An integral surface contains a single infinity of characteristic curves. If then, in space, we are given some twisted curve, not itself a characteristic, and we form the surface which contains the characteristics passing through every point of the given curve, we have an integral surface which satisfies our Cauchy problem. Consider first forming a complete solution of the equation, that is, as we have mentioned above, a solution containing two arbitrary parameters a and b. We shall discuss in the next section how this may be achieved; for the moment we are interested in deriving from a complete integral of (3.6.1):

$$G(x, y, z, a, b) = 0, \tag{3.6.14}$$

an integral surface containing a given curve, defined parametrically by

$$x = x(\sigma), \quad y = y(\sigma), \quad z = z(\sigma). \tag{3.6.15}$$

If for some fixed values of a and b the surface (3.6.14) contains the curve, there is no problem, and so we will assume that it does not. Now suppose that b depends on a, say

$$b = \phi(a).$$

As a varies, the surfaces (3.6.14) will form an envelope, whose equation is obtained in the usual way by eliminating a between

$$G\{x, y, z, a, \phi(a)\} = 0$$

and

$$\frac{\partial G}{\partial a} + \frac{\partial G}{\partial \phi}\phi'(a) = 0.$$

Since the surface G satisfies (3.6.1) for all a and b, so must an envelope of surfaces of this form. Now write

$$\Gamma\{\sigma, a, \phi(a)\} = G\{x(\sigma), y(\sigma), z(\sigma), a, \phi(a)\},$$

so that the envelope will contain the given curve if simultaneously

$$\Gamma\{\sigma, a, \phi(a)\} = 0, \tag{3.6.16}$$

and

$$\frac{\partial \Gamma}{\partial a} + \frac{\partial \Gamma}{\partial \phi}\phi'(a) = 0. \tag{3.6.17}$$

The elimination of the parameter σ between (3.6.16) and (3.6.17) will result in an ordinary first-order equation for $\phi(a)$, which will therefore be determined except for an arbitrary constant. Now (3.6.16) implies that a depends on σ and so differentiating (3.6.16) with respect to σ:

$$\frac{\partial \Gamma}{\partial \sigma} + \left\{\frac{\partial \Gamma}{\partial a} + \frac{\partial \Gamma}{\partial \phi}\phi'(a)\right\}\frac{da}{d\sigma} = 0,$$

or, in the light of (3.6.17)

$$\frac{\partial \Gamma}{\partial \sigma} = 0. \tag{3.6.18}$$

This, however, only contains σ, a, $\phi(a)$, and so the arbitrary constant is fully determined.

It is clear that, independently of Cauchy's problem, we shall always derive a new form of solution of the equation if, in a complete integral, we postulate a relation between the constants $b = \phi(a)$, and form the envelope thus generated. This is referred to as a general solution of the equation. If, as sometimes happens, there is an envelope of the complete integral as a and b vary independently, the resulting envelope will also be an integral surface, and such a surface is referred to as a singular integral.

3.7 Determination of a complete integral

The equation (3.6.1) already gives a single relation between x, y, z, p and q. If we can determine a second relation

$$f(x, y, z, p, q, a), \tag{3.7.1}$$

independent of (3.6.1), and containing an arbitrary constant a, then we may hope to solve (3.6.1) and (3.7.1) to determine p and q as functions of the other variables $p = p(x, y, z, a)$ and $q = q(x, y, z, a)$. These may now be substituted into the usual differential relation for z

$$dz = p(x, y, z, a)\, dx + q\,(x, y, z, a)\, dy \tag{3.7.2}$$

so giving a total differential equation for z, which, if it can be integrated, will introduce a second arbitrary constant b. The result therefore will be a complete integral.

The form of relation (3.7.1) desired will be precisely that which can be obtained as a first integral from the Charpit relations (3.6.13). It will clearly be necessary that such a first integral shall involve either or both of p and q. It is also not trivial that it must be independent of (3.6.1), because (3.6.1) can be obtained as a first integral of the Charpit relations, as can be seen, since

$$\begin{aligned} dF &= X\, dx + Y\, dy + Z\, dz + P\, dp + Q\, dq \\ &\propto \{XP + YQ + Z(pP + qQ) - P(X + pZ) - Q(Y + qZ)\} = 0. \end{aligned}$$

There are a number of special cases which are worth noting.

(*a*) If F contains only p and q then the Charpit relations give

$$dp = dq = 0.$$

Then if $p = a$, $q = q(a)$ is determined by the equation

$$F\{a, q\} = 0,$$

and so

$$dz = a\, dx + q(a)\, dy,$$

and a complete integral is

$$z = ax + q(a)y + b.$$

(*b*) If F contains only p, q and z, the Charpit relations give

$$\frac{dp}{p} = \frac{dq}{q}$$

and so $q = ap$. Taken in conjunction with the original equations this implies

$$p = f(z, a), \quad q = af(z, a),$$

and so

$$\int \frac{dz}{f(z, a)} = x + ay + b.$$

An example is $pq - z = 0$, which leads to

$$\sqrt{ap} = \sqrt{z}, \quad q = \sqrt{(az)},$$

so that

$$4az = (x + ay + b)^2,$$

which represents a family of parabolic cylinders, each of which touches the x, y plane along the length of the generator $x + ay + b = 0$. There is therefore a singular solution $z = 0$, the x, y plane.

It is of interest to notice that in this case the relation $dy = dp$ may also be obtained from the Charpit equations. If then we put $p = y + a$, we derive $q = z/(y + a)$ and

$$dx = \frac{dz}{y + a} - \frac{z\,dy}{(y + a)^2} = d\left\{\frac{z}{y + a}\right\}.$$

Hence another complete integral is

$$z = (x + b)(y + a),$$

a family of hyperbolic paraboloids, also all tangents to the x, y plane. There is therefore nothing unique about a complete integral.

(*c*) If both z and either x or y are missing from the equation, say, for example,

$$F = F(y, p, q),$$

then $X = Z = 0$, and so $p = a$ is an integral of the Charpit relations. This then implies that $q = q(y, a)$, and so a complete integral has the form

$$z = ax + \int q(y, a)\,dy + b.$$

(*d*) If the equation can be written in the form

$$F_1(x, p) = F_2(y, q),$$

then we deduce that

$$dx : dp = \frac{\partial F_1}{\partial p} : -\frac{\partial F_1}{\partial x},$$

so that

$$F_1(x, p) = a = F_2(y, q).$$

These equations may be solved for p and q to give

$$p = \phi_1(x, a), \quad q = \phi_2(y, a)$$

so that a complete integral is

$$z = \int \phi_1(x, a)\, dx + \int \phi_2(y, a)\, dy + b.$$

(*e*) The equation analogous to the Clairaut equation [7], [11] is:

$$z = px + qy + f(p, q).$$

In this case $X = p$, $Y = q$ and $Z = -1$ so that $dp = dq = 0$. A complete integral is therefore

$$z = ax + by + f(a, b).$$

A singular integral will be obtained by eliminating a and b between this and the two equations

$$0 = x + \frac{\partial f}{\partial a} = y + \frac{\partial f}{\partial b}.$$

3.8 A second approach to the Cauchy problem

We may look at the Cauchy problem another way. Equations (3.6.13) are a set of five simultaneous ordinary differential equations, so their solutions can be expressed in terms of some new independent variable t in the form

$$x = x(t, x_0, y_0, z_0, p_0, q_0) \tag{3.8.1}$$

$$y = y(t, x_0, y_0, z_0, p_0, q_0) \tag{3.8.2}$$

$$z = z(t, x_0, y_0, z_0, p_0, q_0) \tag{3.8.3}$$

$$p = p(t, x_0, y_0, z_0, p_0, q_0) \tag{3.8.4}$$

$$q = q(t, x_0, y_0, z_0, p_0, q_0), \tag{3.8.5}$$

where x_0, y_0, z_0, p_0 and q_0 are values taken for $t = 0$, and, as we have seen, only three of these can be chosen independently. The first three of these give the equation of the characteristic, the last two completing the description of the strip. Any integral surface is generated by a one-parameter family of these characteristics, and so a given integral surface will be specified by making the initial values x_0, y_0, z_0, p_0, q_0 depend in some explicit value on a second parameter, say u, so that the quantities above will be functions of the two variables t and u. We must therefore have

$$F\{x, y, z, p, q\} = 0, \tag{3.8.6}$$

$$\frac{\partial z}{\partial t} - p\frac{\partial x}{\partial t} - q\frac{\partial y}{\partial t} = 0, \tag{3.8.7}$$

$$\frac{\partial z}{\partial u} - p\frac{\partial x}{\partial u} - q\frac{\partial y}{\partial u} = 0, \tag{3.8.8}$$

identically in t and u. Now (3.8.1) to (3.8.5) are integrals of the Charpit equation and so imply that $dF = 0$. To satisfy (3.8.6) then it is sufficient that

$$F\{x_0, y_0, z_0, p_0, q_0\} = 0 \tag{3.8.9}$$

identically in u. The Charpit relations ensure also that (3.8.7) is satisfied, and so let

$$H = \frac{\partial z}{\partial u} - p\frac{\partial x}{\partial u} - q\frac{\partial y}{\partial u}.$$

Then

$$\frac{\partial H}{\partial t} = \frac{\partial^2 z}{\partial t\,\partial u} - p\frac{\partial^2 x}{\partial t\,\partial u} - q\frac{\partial^2 y}{\partial t\,\partial u} - \frac{\partial p}{\partial t}\frac{\partial x}{\partial u} - \frac{\partial q}{\partial t}\frac{\partial y}{\partial u}.$$

Now (3.8.7) implies that

$$\frac{\partial^2 z}{\partial t\,\partial u} - p\frac{\partial^2 x}{\partial t\,\partial u} - q\frac{\partial^2 y}{\partial t\,\partial u} - \frac{\partial p}{\partial u}\frac{\partial x}{\partial t} - \frac{\partial q}{\partial u}\frac{\partial y}{\partial t} = 0,$$

and so

$$\frac{\partial H}{\partial t} = \frac{\partial p}{\partial u}\frac{\partial x}{\partial t} - \frac{\partial p}{\partial t}\frac{\partial x}{\partial u} + \frac{\partial q}{\partial u}\frac{\partial y}{\partial t} - \frac{\partial q}{\partial t}\frac{\partial y}{\partial u}$$

or, using the Charpit relations for the t derivatives,

$$\frac{\partial H}{\partial t} = P\frac{\partial p}{\partial u} + X\frac{\partial x}{\partial u} + Q\frac{\partial q}{\partial u} + Y\frac{\partial y}{\partial u} + Z\left\{p\frac{\partial x}{\partial u} + q\frac{\partial y}{\partial u}\right\}. \tag{3.8.10}$$

From (3.8.6), however,

$$X\frac{\partial x}{\partial u} + Y\frac{\partial y}{\partial u} + Z\frac{\partial z}{\partial u} + P\frac{\partial p}{\partial u} + Q\frac{\partial q}{\partial u} = 0,$$

and so

$$\frac{\partial H}{\partial t} = Z\left\{p\frac{\partial x}{\partial u} + q\frac{\partial y}{\partial u} - \frac{\partial z}{\partial u}\right\} = -ZH. \tag{3.8.11}$$

Thus

$$H = H_0 \exp\left\{-\int_0^t Z\,dt\right\}, \tag{3.8.12}$$

and H can only vanish if $H_0 = 0$, i.e. if

$$\frac{\partial z}{\partial u} = p\frac{\partial x}{\partial u} + q\frac{\partial y}{\partial u} \tag{3.8.13}$$

when $t = 0$. This condition with (3.8.9) will determine the surface.

These ideas can then be used to solve the Cauchy problem. The given curve can be assumed to be defined parametrically in the form

$$x = x_0(u), \quad y = y_0(u), \quad z = z_0(u) \tag{3.8.14}$$

and these values can be identified with the initial values for points on the characteristics at $t = 0$. Then (3.8.9) and (3.8.13) will give two equations to determine p and q in terms of u.

Finally therefore the equations (3.8.1) to (3.8.3) will represent the integral surface through the line parametrically in u and t.

3.9 The solution of a Cauchy problem

Let us apply the results of (*d*) of Section 3.7 to show an example of the solution of a Cauchy problem. Consider the problem of finding an integral surface of

$$pq = xy \tag{3.9.1}$$

which contains the curve

$$x = 1, \quad y^3 = z^2. \tag{3.9.2}$$

The theory of Section 3.7 shows that we can take

$$p = ax, \quad q = y/a,$$

and so a complete integral is

$$2az = a^2x^2 + y^2 + b. \tag{3.9.3}$$

Now a variable point on the given curve is $(1, \sigma^2, \sigma^3)$, and so in the notation of Section 3.6:

$$\Gamma\{\sigma, a, \phi(a)\} = 2a\sigma^3 - a^2 - \sigma^4 - \phi(a), \tag{3.9.4}$$

and if we equate this to zero along with

$$\frac{\partial \Gamma}{\partial a} = 2(\sigma^3 - a) - \phi'(a) = 0 \tag{3.9.5}$$

we obtain an equation for $\phi'(a)$, namely

$$[a - \tfrac{1}{2}\{a + \tfrac{1}{2}\phi'(a)\}^{1/3}]\{2a + \phi'(a)\} - \{a^2 + \phi(a)\} = 0, \tag{3.9.6}$$

or if we put $\psi(a) = \frac{1}{2}\{\phi(a) + a^2\}$,

$$[2a - \{\psi'(a)\}^{1/3}]\psi'(a) - 2\psi(a) = 0. \tag{3.9.7}$$

We may also form

$$\frac{\partial \Gamma}{\partial \sigma} = \sigma^2\{6a - 4\sigma\}, \tag{3.9.8}$$

and if this is equated to zero, either $\sigma = 0$ or $\sigma = \frac{3}{2}a$. The insertion of the first into (3.9.4), when the left-hand side is equated to zero, leads to

$$\phi(a) + a^2 = 0,$$

which is clearly a solution of (3.9.7). If we put this into (3.9.3) we derive the surface

$$2az + a^2(1 - x^2) - y^2 = 0, \tag{3.9.9}$$

and the envelope of this, as a varies, is obtained by eliminating a between (3.9.9) and

$$z + a(1 - x^2) = 0. \tag{3.9.10}$$

The envelope therefore has the equation

$$z^2 = (x^2 - 1)y^2 \tag{3.9.11}$$

which manifestly does not contain (3.9.2).

On the other hand, if $\sigma = \frac{3}{2}a$, (3.9.4) leads to

$$\phi(a) + a^2 = \tfrac{27}{16}a^4 \tag{3.9.12}$$

which is also a solution of (3.9.7). If this is inserted into (3.9.3) we get

$$2az + a^2(1 - x^2) - y^2 - \tfrac{7}{16}a^4 = 0, \tag{3.9.13}$$

and combining this with

$$z + a(1 - x^2) - \tfrac{27}{8}a^3 = 0 \tag{3.9.14}$$

a surface of high order will be obtained. It is plain, however, from (3.9.13) and (3.9.14) that the curve in which this cuts $x = 1$ is given parametrically by

$$y = (\tfrac{3}{2}a)^2, \quad z = (\tfrac{3}{2}a)^3,$$

and so the surface does contain the given curve.

Alternatively, if we follow Section 3.8 the solution of the Charpit equations may be written

$$\begin{aligned}
x &= x_0 \cosh t + q_0 \sinh t,\\
y &= y_0 \cosh t + p_0 \sinh t,\\
z &= z_0 + \tfrac{1}{2}(y_0q_0 + x_0p_0)(\cosh 2t - 1) + \tfrac{1}{2}(p_0q_0 + x_0y_0)\sinh 2t,\\
p &= y_0 \sinh t + p_0 \cosh t,
\end{aligned}$$

and $q = x_0 \sinh t + q_0 \cosh t$.

We may, as before, represent a point on the given curve by the triad

$$(1, u^2, u^3),$$

and so the equations corresponding to (3.8.9) and (3.8.13) are

$$p_0q_0 = u^2 \tag{3.9.15}$$

and

$$3u^2 = 2q_0u. \tag{3.9.16}$$

Thus the integral surface is represented parametrically by

$$x(t, u) = \cosh t + \tfrac{3}{2}u \sinh t, \tag{3.9.17}$$

$$y(t, u) = u^2 \cosh t + \tfrac{2}{3}u \sinh t, \tag{3.9.18}$$

and

$$z(t, u) = u^3 + \tfrac{1}{2}(\tfrac{3}{2}u^3 + \tfrac{2}{3}u)(\cosh 2t - 1) + u^2 \sinh 2t. \tag{3.9.19}$$

3.10 The one-dimensional Hamilton-Jacobi equation

A rather trivial example of the use of the theory above is to apply it to the Hamilton-Jacobi equation for the one-dimensional motion of a single particle in a conservative field of force. The example does, nevertheless, give some insight into the theory that has been developed.

If the particle has mass m, and the potential of the field is $V(x)$, the generalized momentum is $p = m\dot{x}$, and so the Hamiltonian H is given by

$$2mH(p, x) = p^2 + 2mV(x). \tag{3.10.1}$$

Hence if we put $q = \partial S/\partial t$, and now interpret p as $\partial S/\partial x$ the Hamilton-Jacobi equation is:

$$p^2 + 2mq + 2mV(x) = 0. \tag{3.10.2}$$

The Charpit equations are

$$\frac{m\,dx}{p} = \frac{dt}{1} = \frac{m\,dS}{p^2 + mq} = -\frac{dp}{V'(x)} = -\frac{dq}{0}. \tag{3.10.3}$$

From these one can deduce immediately $q =$ const., the usual energy integral, and the pair of ordinary differential equations

$$\frac{m\,dx}{dt} = p \tag{3.10.4}$$

and

$$\frac{dp}{dt} = -V'(x), \tag{3.10.5}$$

which are obviously just the ordinary equations of motion. This therefore shows the trajectories obtained by solving the equations of motion and also using the remaining relation

$$m\frac{dS}{dt} = p^2 + mq \tag{3.10.6}$$

are characteristics in the (x, t, s) space. If, further, we notice that the equation (3.10.2) is of the type (d) of Section 3.7, we can obtain a complete integral of (3.10.2) in the form

$$S(x, t, a, b) = \int \{2m(a - V\}^{1/2}\,dx - at + b. \tag{3.10.7}$$

Then we notice that the significance of (1.2.14) is that

$$\frac{\partial S}{\partial x} = \{2m(a - V)\}^{1/2} = p \tag{3.10.8}$$

which was otherwise known, and

$$\frac{\partial S}{\partial a} = m^{1/2}\{a - V(x)\}^{-1/2} - t, \tag{3.10.9}$$

and an integral of the equation of motion is obtained by equating this to a constant

$$m^{1/2}\{a - V(x)\}^{-1/2} - t = c. \tag{3.10.10}$$

3.11 The first-order equation in n independent variables

This time we work in the $(n + 1)$-dimensional space with coordinates x_i, $i = 1, 2, \ldots, n$ and z, and as usual we treat x_i as the ith component of the vector $\mathbf{x}$, and $p_i = \partial z/\partial x_i$ as the ith component of a vector $\mathbf{p}$. Thus the equation will take the form

$$F\{\mathbf{x}, z, \mathbf{p}\} = 0. \tag{3.11.1}$$

At a fixed point of the space $(\mathbf{x}, z)$, therefore, the vector $\mathbf{p}$ has $n - 1$ degrees of freedom. If, following the previous notation, we put

$$P_i = \frac{\partial F}{\partial p_i}, \quad X_i = \frac{\partial F}{\partial x_i} \quad \text{and} \quad Z = \frac{\partial F}{\partial z},$$

then it is clear that small changes δp_i in the values of p_i at the given point must satisfy

$$\sum_{i=1}^{n} P_i \, \delta p_i = 0. \tag{3.11.2}$$

A tangent plane at $(\mathbf{x}, z)$ to an integral surface of (3.11.1) has the equation

$$\zeta - z = \sum_{i=1}^{n} p_i(\xi_i - x_i). \tag{3.11.3}$$

This will intersect the neighbouring tangent plane

$$\zeta - z = \sum_{i=1}^{n} (p_i + \delta p_i)(\xi_i - x_i) \tag{3.11.4}$$

in the hyperplane

$$\sum_{i=1}^{n} \delta p_i(\xi_i - x_i) = 0. \tag{3.11.5}$$

If we use the Lagrange undetermined multiplier λ to combine (3.11.2) and (3.11.5), then

$$\sum_{i=1}^{n} \delta p_i\{\lambda P_i - (\xi_i - x_i)\} = 0, \tag{3.11.6}$$

and we can now treat the δp_i as independent, but determining λ. Thus

$$\lambda = \frac{\xi_i - x_i}{P_i} \quad \text{for} \quad i = 1, 2, \ldots, n. \tag{3.11.7}$$

This relation is clearly analogous to the one derived in Section 3.6. If we define the characteristic as a curve whose tangent is everywhere directed along a generator of the envelope of the tangent planes at $(\mathbf{x}, z)$ to the integral surfaces, then as in Section 3.6 we can derive the equations of the characteristics by letting $\xi_i \to x_i$ in (3.11.7). Thus the equations of the characteristics satisfy

$$\frac{dx_i}{P_i} = \frac{\sum_{i=1}^{n} p_i\, dx_i}{\sum_{i=1}^{n} p_i P_i} = \frac{dz}{\sum_{i=1}^{n} p_i P_i} = d\lambda \quad i = 1, 2, \ldots, n. \tag{3.11.8}$$

Now write $\partial^2 z/\partial x_i\, \partial x_j = p_{ij}$, so that, if we differentiate (3.11.1) with respect to x_j, we derive

$$X_i + p_i Z + \sum_{j=1}^{n} p_{ij} P_j = 0. \tag{3.11.9}$$

We know also that

$$dp_i = \sum_{j=1}^{n} p_{ij}\, dx_j = \left\{ \sum_{j=1}^{n} p_{ij} P_j \right\} d\lambda$$

if we use (3.11.8). Finally, therefore, we derive a complete set of relations which generalize those of Charpit:

$$\frac{dx_i}{P_i} = \frac{dz}{\sum_{i=1}^{n} p_i P_i} = -\frac{dp_i}{X_i + p_i Z} \quad i = 1, 2, \ldots, n. \tag{3.11.10}$$

We can generalize also the analysis of Section 3.8 to solve the Cauchy problem. In fact, certain difficulties can arise in the analysis, but to discuss these would take us beyond the scope of this book. For a fuller discussion the reader is recommended to the appropriate chapter of [7].

A solution of the equation (3.11.10) can be expressed in terms of an auxiliary variable t, and the $2n + 1$ constants which represent the values of $\mathbf{x}$, z and $\mathbf{p}$ when $t = 0$. On the other hand, an integral of (3.11.1) is an n-dimensional hypersurface and so contains ∞^{n-1} of the characteristic curves (3.11.10). Thus the $2n + 1$ constants can be made to depend on $n - 1$ auxiliary variables $u_1, u_2, \ldots, u_{n-1}$. As in the two-variable case in Section 3.8 we must have as an identity in the u_i

$$F\{\mathbf{x}_0, z_0, \mathbf{p}_0\} = 0. \tag{3.11.11}$$

We also know that

$$\frac{\partial z}{\partial t} = \sum_{i=1}^{n} p_i \frac{\partial x_i}{\partial t}, \tag{3.11.12}$$

and if δx_i represents a displacement of x_i due to arbitrary small changes in the u_i, then identically

$$\delta z = \sum_{i=1}^{n} p_i\, \delta x_i. \tag{3.11.13}$$

Now let

$$H = \delta z - \sum_{i=1}^{n} p_i\, \delta x_i,$$

then

$$\frac{\partial H}{\partial t} = \frac{\partial}{\partial t}(\delta z) - \sum_{i=1}^{n} \frac{\partial p_i}{\partial t}\delta x_i - \sum_{i=1}^{n} p_i \frac{\partial}{\partial t}(\delta x_i).$$

But $\partial/\partial t$ and δ are commutative operators, and

$$\delta\left(\frac{\partial z}{\partial t}\right) = \sum_{i=1}^{n} \delta p_i \frac{\partial x_i}{\partial t} + \sum_{i=1}^{n} p_i \delta\left(\frac{\partial x_i}{\partial t}\right),$$

so that

$$\frac{\partial H}{\partial t} = \sum_{i=1}^{n}\left(\delta p_i \frac{\partial x_i}{\partial t} - \frac{\partial p_i}{\partial t}\delta x_i\right)$$

$$= \sum_{i=1}^{n} \{P_i\, \delta p_i + (X_i + p_i Z)\, \delta x_i\},$$

from the Charpit relations. Since, identically, $F\{x, y, z, p, q\} = 0$, it follows that

$$\sum_{i=1}^{n} (P_i\, \delta p_i + X_i\, \delta x_i) + Z\, \delta z = 0,$$

and so

$$\frac{\partial H}{\partial t} = -ZH. \qquad (3.11.14)$$

Once again therefore

$$H = H_0 \exp\left(-\int_0^t Z\, dt\right),$$

and so (3.11.13) implies that $H_0 = 0$. Thus we deduce a further identity in the u_i, that

$$\delta z_0 - \sum_{i=1}^{n} p_{i0}\, \delta x_{i0} = 0. \qquad (3.11.15)$$

In particular, as in Section 3.8, the Cauchy problem consists of finding the n-dimensional hypersurface generated by the characteristics which pass through a given piece of $(n - 1)$-dimensional hypersurface. If then the given hypersurface is parametrized, the $n - 1$ parameters needed can be taken to be the u_i above, and the problem is solved.

3.12 An example of a Cauchy problem in four variables

Let us consider the equation

$$p_1 p_2 + p_2 p_3 + p_3 p_1 = x_1 x_2 + x_2 x_3 + x_3 x_1. \qquad (3.12.1)$$

From (3.11.8) the equations for the characteristic strip may be written

$$\frac{dx_i}{dt} = \pi - p_i, \quad i = 1, 2, 3, \tag{3.12.2}$$

$$\frac{dp_i}{dt} = -\xi + x_i, \quad i = 1, 2, 3, \tag{3.12.3}$$

$$\frac{dz}{dt} = 2(x_1x_2 + x_2x_3 + x_3x_1), \tag{3.12.4}$$

where we have written $\xi = x_1 + x_2 + x_3$, $\pi = p_1 + p_2 + p_3$.

Therefore
$$\frac{d^2\xi}{dt^2} = 2\frac{d\pi}{dt} = -4\xi, \tag{3.12.5}$$

and so

$$\xi = (x_{10} + x_{20} + x_{30}) \cos 2t + (p_{10} + p_{20} + p_{30}) \sin 2t, \tag{3.12.6}$$

$$\pi = (p_{10} + p_{20} + p_{30}) \cos 2t - (x_{10} + x_{20} + x_{30}) \sin 2t. \tag{3.12.7}$$

It can also be seen that

$$\frac{d^2}{dt^2}\{x_1 - x_2\} = \frac{d}{dt}\{p_2 - p_1\} = x_2 - x_1, \tag{3.12.8}$$

and a similar relation can be derived for, say, $x_1 - x_3$.

It is clear that if we know the expressions for one of the x_i and one of the p_i the others may be obtained by cyclic permutation of the suffices. Then, for example,

$$\begin{aligned} 3x_1 &= x_{10}(\cos 2t + 2 \cos t) + (x_{20} + x_{30})(\cos 2t - \cos t) \\ &\quad + p_{10}(\sin 2t - 2 \sin t) + (p_{20} + p_{30})(\sin 2t + \sin t) \end{aligned} \tag{3.12.9}$$

$$\begin{aligned} 3p_1 &= p_{10}(\cos 2t + 2 \cos t) + (p_{20} + p_{30})(\cos 2t - \cos t) \\ &\quad - x_{10}(\sin 2t - 2 \sin t) - (x_{20} + x_{30})(\sin 2t + \sin t). \end{aligned} \tag{3.12.10}$$

We may find a closed expression for z in the following way. We notice that

$$\begin{aligned} \frac{d}{dt}(x_1p_1 + x_2p_2 + x_3p_3) &= 2(p_2p_3 + p_3p_1 + p_1p_2) \\ &\quad - 2(x_2x_3 + x_3x_1 + x_1x_2) = 0, \end{aligned}$$

and so

$$x_1p_1 + x_2p_2 + x_3p_3 = x_{10}p_{10} + x_{20}p_{20} + x_{30}p_{30} = \tau_0, \text{ say.} \tag{3.12.11}$$

Furthermore

$$\frac{d}{dt}(x_2x_3 + x_3x_1 + x_1x_2) = \xi\pi + x_1p_1 + x_2p_2 + x_3p_3,$$

and so

$$\begin{aligned} \frac{dz}{dt} &= x_{20}x_{30} + x_{30}x_{10} + x_{10}x_{20} + \tfrac{1}{4}\xi_0\pi_0 \sin 4t \\ &\quad + \tfrac{1}{8}(\pi_0^2 - \xi_0^2)(1 - \cos 4t) + \tau_0, \end{aligned}$$

and

$$z = z_0 + \{x_{20}x_{30} + x_{30}x_{10} + x_{10}x_{20} + \tau_0 + \tfrac{1}{8}(\pi_0^2 - \xi_0^2)\}t + \frac{\xi_0\pi_0}{16}\{1 - \cos 4t\} - \frac{\pi_0^2 - \xi_0^2}{32}\sin 4t. \quad (3.12.12)$$

Suppose now we want the solution which contains the two-dimensional surface $z = 1$, $x_1^2 + x_2^2 + x_3^2 = 1$. We can represent this initial surface parametrically by the tetrad

$$[\cos u, \sin u \cos v, \sin u \sin v, 1].$$

These values can be substituted for x_{10}, x_{20}, x_{30} and z_0 in the equations above. To determine the appropriate initial values of the p_i we have from (3.12.1)

$$p_{10}p_{20} + p_{20}p_{30} + p_{30}p_{10} = \sin u \cos u\{\sin v + \cos v\} + \sin^2 u \sin v \cos v. \quad (3.12.13)$$

Equation (3.11.15), written out in full, is

$$\delta z_0 = p_{10}\,\delta x_{10} + p_{20}\,\delta x_{20} + p_{30}\,\delta x_{30},$$

or identically in u and v and for all δu and δv:

$$0 = \{-p_{10}\sin u + p_{20}\cos u \cos v + p_{30}\cos u \sin v\}\,\delta u + \sin u\{-p_{20}\sin v + p_{10}\cos v\}\,\delta v.$$

Thus $p_{10} = x_{10}$, $p_{20} = x_{20}$ and $p_{30} = x_{30}$, and when these values are used in (3.12.9) and (3.12.12) we derive a representation of x_1, x_2, x_3, z in terms of the three parameters t, u and v. This is a three-dimensional hypersurface in the four-dimensional space, it is an integral surface of equation (3.12.1), and it contains the given two-dimensional surface. It is therefore a solution of the Cauchy problem.

Exercises

3.1. Find the family of characteristics of the equations

(*a*) $\dfrac{\partial u}{\partial x} + x\dfrac{\partial u}{\partial y} = 0$

(*b*) $y\dfrac{\partial u}{\partial x} + x\dfrac{\partial u}{\partial y} = 0$

(*c*) $2xy\dfrac{\partial u}{\partial x} + (y^2 - x^2 + 1)\dfrac{\partial u}{\partial y} = 0.$

3.2. Find the solution of 3.1 (*a*) on the half plane $x > 0$ given that on $x = 0$, $u(0, y) = \sin y$.

3.3. Show that the solution of 3.1 (*c*) is determined in the region $x > 0, y > 0$, if u is given on $y = 0$, $0 \leqslant x \leqslant 1$. Find this solution if $u(x, 0) = x(1 - x)$, $0 \leqslant x \leqslant 1$.

3.4. Solve the equation

$$y\frac{\partial u}{\partial x} + x\frac{\partial u}{\partial y} = xy$$

in the region $x > 0$, $y > 0$ given that on $x = 0$, $y > 0$, $u(0, y) = \exp(-y^2)$ and on $y = 0$, $x > 0$, $u(x, 0) = \exp(-x^2)$.

3.5. Find the surface generated by the characteristics of

$$yz\frac{\partial u}{\partial x} + zx\frac{\partial u}{\partial y} + xy\frac{\partial u}{\partial z} = 0$$

which pass through the circle $x = 0$, $y^2 + z^2 = a^2$.

3.6. Find a complete integral of the equations

(*a*) $p^2 + q^2 = 1$
(*b*) $p^2 + q^2 = z$
(*c*) $y^2p^2 + q^2 = 1$
(*d*) $x^2p^2 + y^2q^2 = 1$
(*e*) $x^2p^2 + y^2q^2 = z$.

3.7. By using the method of envelopes of complete integrals find an integral surface of $p^2 + q^2 = z$ (Example 3.6(*b*) above) which contains the curve $z = 1$, $x^2 + y^2 = 1$.

3.8. Repeat the above example but using the characteristic strip method, instead of the envelope method.

3.9. Compare the two methods for finding the integral surface of the equation of example 3.6(*c*) above, which contains the curve $y = 0$, $x^2 + z^2 = 1$.

Answers

3.1. (*a*) The parabolae $y = x^2 + c$, (*b*) the hyperbolae $x^2 - y^2 - c$, (*c*) the circles $x^2 - y^2 - 2cx + 1 = 0$, $c > 1$.

3.2. $u(x, y) = \sin(y - x^2)$.

3.3. $u(x, y) = 2(v^2 - 1)^{1/2}\{v - (v^2 - 1)^{1/2}\}^2$,
where $v(x, y) = (x^2 - y^2 + 1)/2x$.

3.4. $u(x, y) = \exp\{-|x^2 - y^2|\}$.

3.5. $y^2 + z^2 - 2x^2 = a^2$.

3.6. There are no unique complete integrals, but if the standard methods are applied to (*a*)–(*d*) the answers are

(*a*) $z = x \sin a + y \cos a + b$
(*b*) $4(1 + a^2)z = (x + ay + b)^2$
(*c*) $az = a^2x + \sin^{-1} ay + b$
(*d*) $z = \sin a \log x + \cos a \log y + b$.
(*e*) A complete integral is

$$z(a + \tfrac{1}{2}\log x)^2 + (\log by)^2.$$

3.7 (and 3.8). $(x^2 + y^2 - 4z^2 - 1)^2 = 16z$.

CHAPTER 4

Second-order Linear Equations

4.1 Simultaneous first-order equations

We shall first consider equations in two independent variables only. As in the case of first-order equations, we call a second-order partial differential equation linear if it can be written as:

$$A(x, y)\Phi_{xx} + 2B(x, y)\Phi_{xy} + C(x, y)\Phi_{yy} + E(x, y)\Phi_x + F(x, y)\Phi_y + G(x, y)\Phi = D(x, y), \quad (4.1.1)$$

where suffices denote partial differentiations. If it can be written only as

$$A(x, y)\Phi_{xx} + 2B(x, y)\Phi_{xy} + C(x, y)\Phi_{yy} = D(x, y, \Phi, \Phi_x, \Phi_y), \quad (4.1.2)$$

we call it semi-linear, and if all the coefficients A, B, C and D depend not only on x and y, but also on Φ, Φ_x and Φ_y, we call the equation quasi-linear.

For a number of purposes it is convenient to consider alternatively a pair of simultaneous first-order equations. The analogous system to (4.1.2), which therefore would be classified as semi-linear, would be

$$a_{11}(x, y)u_x + a_{12}(x, y)u_y + b_{11}(x, y)v_x + b_{12}(x, y)v_y = c_1\{x, y, u, v\}, \quad (4.1.3)$$

and

$$a_{21}(x, y)u_x + a_{22}(x, y)u_y + b_{21}(x, y)v_x + b_{22}(x, y)v_y = c_2\{x, y, u, v\}. \quad (4.1.4)$$

It may reasonably be objected that these two systems are not identical, for if in (4.1.2) we write $u = \Phi_x$, $v = \Phi_y$, we derive the equation pair

$$A(x, y)u_x + B(x, y)u_y + B(x, y)v_x + C(x, y)v_y = D(x, y, \Phi, u, v), \quad (4.1.5)$$

and

$$u_y = v_x, \quad (4.1.6)$$

which are comparable with (4.1.3) and (4.1.4) only if Φ does not appear explicitly on the right-hand side of (4.1.5).

On the other hand, provided either

$$a_{11}a_{22} - a_{12}a_{21} \neq 0 \text{ or } b_{11}b_{22} - b_{12}b_{21} \neq 0$$

we can solve (4.1.3) and (4.1.4) as a pair of simultaneous equations to determine either u_x and u_y or v_x and v_y. If, for example,

$$b_{11}b_{22} - b_{12}b_{21} \neq 0,$$

we would derive relations of the form

$$v_x = \alpha_1(x, y)u_x + \beta_1(x, y)u_y + \gamma_1(x, y, u, v), \qquad (4.1.7)$$

and

$$v_y = \alpha_2(x, y)u_x + \beta_2(x, y)u_y + \gamma_2(x, y, u, v). \qquad (4.1.8)$$

Provided that we may assume the equivalence of v_{xy} and v_{yx} we may use this to derive a single second-order equation for u:

$$\alpha_2(x, y)u_{xx} + \{\beta_2(x, y) - \alpha_1(x, y)\}u_{xy} - \beta_1(x, y)u_{yy} = \Gamma(x, y, u, v, u_x, u_y, v_x, v_y). \qquad (4.1.9)$$

We may again use (4.1.7) and (4.1.8) to eliminate v_x and v_y from the right-hand side of (4.1.9), but it will still, in general, be a function of v, and so (4.1.9) and (4.1.2) are not identical.

In each case we see therefore that the difference between (4.1.2) and the equation pair (4.1.3) and (4.1.4) is the presence, as an argument, of a function, Φ in the one case, v in the other, which, although not directly determined by the equation, has both its first partial derivatives so determined. Now a function whose first partial derivatives are everywhere known and are consistent is given uniquely except for an arbitrary numerical constant. As long as a single numerical value of the function is known therefore, this ambiguity is removed, and the two systems may be considered identical.

4.2 Classification of linear equations of second order in two independent variables

If

$$\frac{a_{11}(x, y)}{b_{11}(x, y)} = \frac{a_{12}(x, y)}{b_{12}(x, y)} = k(x, y) \qquad (4.2.1)$$

then (4.1.3) can, by an argument similar to that which we employed for first-order equations, be regarded as a total differential relation

$$\left\{k(x, y)\frac{du}{ds} + \frac{dv}{ds}\right\}[\{b_{11}(x, y)\}^2 + \{b_{12}(x, y)\}^2]^{1/2} = c_1\{x, y, u, v\}, \qquad (4.2.2)$$

where ds is an increment of arc length, along a curve which is an integral curve of the equation

$$b_{11}(x, y)\frac{dy}{dx} = b_{12}(x, y). \qquad (4.2.3)$$

Similarly if it happens that the coefficients of (4.1.4) are in a similar ratio, we can regard this equation as a total differential relation in a particular direction. In general, of course, no such simple relation would be exhibited by the coefficients of either equation.

If we take a linear combination of the two equations

$$(\lambda_1 a_{11} + \lambda_2 a_{21})u_x + (\lambda_1 a_{12} + \lambda_2 a_{22})u_y + (\lambda_1 b_{11} + \lambda_2 b_{21})v_x + (\lambda_1 b_{12} + \lambda_2 b_{22})v_y = \lambda_1 c_1 + \lambda_2 c_2 \quad (4.2.4)$$

for suitable values of λ_1 and λ_2, it may be possible that

$$\frac{\lambda_1 a_{11} + \lambda_2 a_{21}}{\lambda_1 b_{11} + \lambda_2 b_{21}} = \frac{\lambda_1 a_{12} + \lambda_2 a_{22}}{\lambda_1 b_{12} + \lambda_2 b_{22}} = k(x, y), \quad (4.2.5)$$

in which case (4.2.4) would reduce to the total differential equation

$$k(x, y)\frac{du}{ds} + \frac{dv}{ds} = \gamma(x, y, u, v), \quad (4.2.6)$$

where the differentiation is along an integral curve of

$$(\lambda_1 b_{11} + \lambda_2 b_{21})\frac{dy}{dx} = \lambda_1 b_{12} + \lambda_2 b_{22}, \quad (4.2.7)$$

and we have put

$$\{(\lambda_1 b_{11} + \lambda_2 b_{21})^2 + (\lambda_1 b_{12} + \lambda_2 b_{22})^2\}^{1/2}\gamma = \lambda_1 c_1 + \lambda_2 c_2.$$

The combination of (4.2.5) and (4.2.7) would imply that

$$\zeta = \frac{dy}{dx} = \frac{\lambda_1 a_{12} + \lambda_2 a_{22}}{\lambda_1 a_{11} + \lambda_2 a_{21}} = \frac{\lambda_1 b_{12} + \lambda_2 b_{22}}{\lambda_1 b_{11} + \lambda_2 b_{21}},$$

or

$$(a_{12} - a_{11}\zeta)\lambda_1 + (a_{22} - a_{21}\zeta)\lambda_2 = 0,$$

and

$$(b_{12} - b_{11}\zeta)\lambda_1 + (b_{22} - b_{21}\zeta)\lambda_2 = 0,$$

a pair of homogeneous simultaneous equations for λ_1 and λ_2. Since the equations are homogeneous, either

$$\lambda_1 = \lambda_2 = 0,$$

or

$$\begin{vmatrix} a_{12} - a_{11}\zeta & a_{22} - a_{21}\zeta \\ b_{12} - b_{11}\zeta & b_{22} - b_{21}\zeta \end{vmatrix} = 0. \quad (4.2.8)$$

This determinantal equation expands to

$$d_1 - (d_2 + d_3)\zeta + d_4\zeta^2 = 0, \quad (4.2.9)$$

where $d_1 = a_{12}b_{22} - a_{22}b_{12}$, $d_2 = a_{12}b_{21} - a_{21}b_{12}$, $d_3 = a_{11}b_{22} - a_{22}b_{11}$, and $d_4 = a_{11}b_{21} - a_{21}b_{11}$.

Since the coefficients of the quadratic (4.2.9) are real, its roots may be real and distinct, real and coincident or conjugate complex. Corresponding to these three cases we say that the equation pair (4.1.3) and (4.1.4) is hyperbolic, parabolic or elliptic.

If then we are considering the second-order equation (4.1.2), we can replace it by the equation pair (4.1.5) and (4.1.6), and proceed as above. It can then be easily seen that the quadratic equivalent to (4.2.9) is:

$$C(x, y) - 2B(x, y)\zeta + A(x, y)\zeta^2 = 0.$$

4.3 Illustrative examples

We will consider some examples to illustrate the classification.

(*a*) The two-variable Laplace equation $\Phi_{xx} + \Phi_{yy} = 0$. The corresponding quadratic equation is $\zeta^2 + 1 = 0$, so that the equation is elliptic.

(*b*) The two-variable wave equation $\Phi_{xx} - \Phi_{yy} = 0$. The corresponding quadratic is $\zeta^2 - 1 = 0$, and so the equation is hyperbolic.

(*c*) The two-variable diffusion equation $\Phi_{xx} - \Phi_y = 0$, which leads to $\zeta^2 = 0$ and so is parabolic.

We consider now two examples of slightly greater substance.

(*d*) The Tricomi equation $\Phi_{xx} - x\Phi_{yy} = 0$, which is of interest as an approximation to the equation for the velocity potential of an irrotational two-dimensional gas flow near to the velocity of sound. In this case the quadratic is

$$\zeta^2 - x = 0.$$

The original equation is therefore elliptic in the half-plane $x < 0$, and hyperbolic in the half-plane $x > 0$. In the hyperbolic half-plane we shall have as the roots of the quadratic

$$\frac{dy}{dx} = \pm x^{1/2}$$

which integrate to give

$$y = C \pm \tfrac{2}{3}x^{3/2},$$

families of semi-cubical parabolae.

(*e*) The equation

$$(x^2 - 1)\Phi_{xx} + 2xy\Phi_{xy} + (y^2 - 1)\Phi_{yy} + 2x\Phi_x + 2y\Phi_y = 0.$$

This equation again is of some physical interest, arising as it does in the theory of linearized cone fields in aerodynamics.

The corresponding quadratic is

$$(x^2 - 1)\zeta^2 - 2xy\zeta + (y^2 - 1) = 0,$$

and this has real roots if

$$x^2y^2 \geqslant (x^2 - 1)(y^2 - 1)$$

or

$$x^2 + y^2 \geqslant 1.$$

This equation therefore is elliptic within the unit circle, and hyperbolic outside it. In the hyperbolic case, the differential equation for the characteristics can be rewritten as

$$\left(x\frac{dy}{dx} - y\right)^2 = 1 + \left(\frac{dy}{dx}\right)^2$$

or

$$y = x\frac{dy}{dx} \pm \left\{1 + \left(\frac{dy}{dx}\right)^2\right\}^{1/2},$$

which is a Clairaut differential equation (see Ref. [7] or [11]).

Hence the integral curves are the lines

$$y = mx \pm (1 + m^2)^{1/2},$$

all tangents to the unit circle, which is their envelope. The unit circle itself also forms a singular integral curve of the differential equation.

4.4 The hyperbolic equation

We return now to the quadratic (4.2.9) and we shall assume for the moment that we have to do with a hyperbolic equation, so that the roots of the quadratic are real and distinct. Corresponding to the sign of the square root which is taken, we can distinguish the two roots as ζ^+ and ζ^-. We shall then derive, as we have seen in the examples just considered, two first-order ordinary differential equations

$$\frac{dy}{dx} = \zeta^+ \text{ and } \frac{dy}{dx} = \zeta^-. \tag{4.4.1}$$

With the usual provisos, each of these can be integrated to give a one-parameter family of integral curves, and we call these the characteristics of the original equation. Thus we see that in that region of the plane in which an equation is hyperbolic, any given point will have two characteristic curves passing through it, one member of each of the families. We can therefore say that a hyperbolic equation has two families of real characteristics, a parabolic equation has one such family, taken twice over, and an elliptic equation has complex characteristics.

If the root ζ^+ is taken, then the first of the simultaneous equations will give a unique value for the ratio λ_1/λ_2. Corresponding to this ratio (4.2.5) will define a function $k^+(x, y)$ so that

$$(b_{11}a_{22} - b_{21}a_{12} + d_4\zeta^+)k^+(x, y) = a_{11}a_{22} - a_{21}a_{12} \tag{4.4.2}$$

and a function $\gamma^+(x, y)$ for which

$$\{(1 + \zeta^{+2})^{1/2}(a_{11}a_{22} - a_{21}a_{12})\}\gamma^+(x, y) = \{c_1(a_{22} - a_{21}\zeta^+) - c_2(a_{12} - a_{11}\zeta^+)\}k^+(x, y). \tag{4.4.3}$$

With these values substituted into it, (4.2.6) becomes a total differential equation, since from our knowledge of the integral curve of (4.4.1) we can regard both x and y as functions of the arc length s. In principle therefore, (4.2.6) can, possibly with a suitable integrating factor, be integrated along the characteristic to give a relation between the values that u and v take at each point of the characteristic. We can derive similar relations on the other characteristic $dy/dx = \zeta^-$; clearly all we need do in (4.4.2) and (4.4.3) is to change the $^+$ superfixes to $^-$, and a second relation between u and v can be found which will hold on each of the second family of characteristics. We will use the previous examples to illustrate this.

A Cauchy problem would be given if we knew not only the pair of

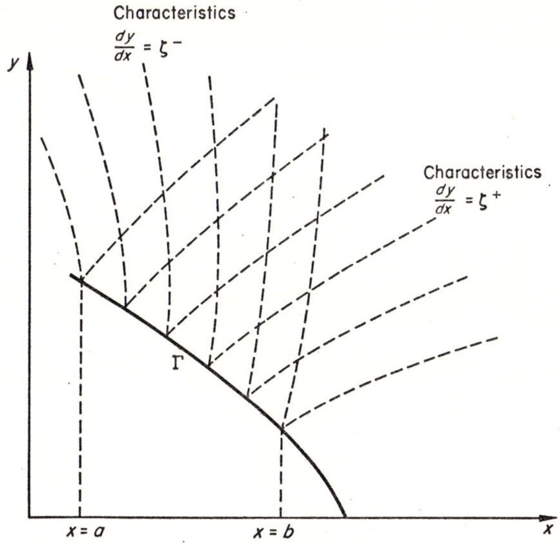

Fig. 4.1

equations (4.1.3) and (4.1.4), but also the distribution of values of u and v on a segment of some curve Γ (see Fig. 4.1). If Γ is itself a part of one of the characteristics of either family, the u and v cannot be given arbitrary values on Γ, but must satisfy the relation between u and v which is appropriate to that characteristic, and the Cauchy problem is indeterminate. If Γ is not characteristic, then the values of u and v can be specified arbitrarily. The relation between u and v on a ζ^+ characteristic, being the integral of a total differential equation, will contain a constant of integration, and so for any ζ^+ characteristic intersecting Γ the constant can be found from the initial data. We can therefore assume that u is given as a single-valued function of v on each ζ^+ characteristic intersecting Γ. Similarly we can find a second relation between u and v

on each ζ^- characteristic intersecting Γ. In that region of the plane which is covered by both sets of the characteristics which also cut Γ, two known and distinct relations between u and v must hold, and so u and v are determined. The Cauchy problem is therefore solved.

Conversely, if through any point of the plane we draw the pair of characteristics which pass through it, we can see that the values of u and v at the point will be influenced only by the initial data given on those segments of the curves which lie between the characteristics. This means of course that, in a particular problem, since the value of u and v at P (see Fig. 4.2) depends only on initial data given on the part of Γ

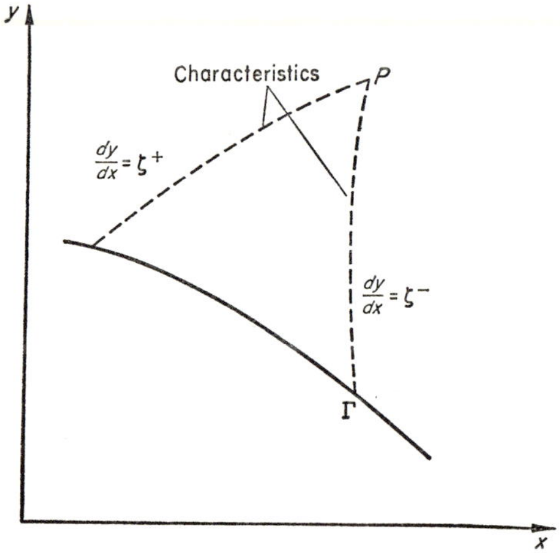

Fig. 4.2

intercepted between the characteristics, the data outside could be changed, without altering the value at P. This leads us to the conclusion that we can join different solutions of the partial differential equation along a characteristic, and the discontinuity will 'propagate' along the characteristic. It is indeed true that in most hyperbolic problems of mathematical physics the characteristics can be interpreted as wavelets, with all this implies to the physical intuition.

We can see this idea of the 'patching' of different solutions from a slightly different viewpoint. Suppose that we refer equations (4.1.3) and (4.1.4) to a system of orthogonal curvilinear coordinates. Generally the equations will be of the same sort as we have at the moment. However, in the particular case where we base our coordinate system on one of the families of characteristics and its orthogonal trajectories, the fact that (4.2.6) is a total differential equation means that the only partial derivatives in our transformed equations will be along the characteristics, and no normal derivatives will appear. Joining a new solution on at a characteristic will not then violate the original equation.

We remark finally that since discontinuities can occur in the boundary data without any serious consequences for the solution, there is no need for the initial data to be analytic as the Cauchy-Kowalewski existence theorem assumed.

4.5 Further discussion of the examples

We revert now to those of our previous examples (Section 4.3) which gave hyperbolic equations.

(*b*) We interpret this now as the equation pair

$$\left.\begin{aligned} u_x &= v_y, \\ u_y &= v_x. \end{aligned}\right\} \tag{4.5.1}$$

The roots were $\zeta^+ = 1$ and $\zeta^- = -1$, so that (4.4.2) and (4.4.3) just give $k^+(x, y) = -1$, $k^-(x, y) = 1$, $\gamma^+ = \gamma^- = 0$.

Corresponding then to (4.2.6), we find that on the line $dy/dx = 1$ or $y - x = c$, $u - v$ is constant, and on $dy/dx = -1$ or $y + x = c$, $u + v$ is constant.

(*d*) Interpreting the Tricomi equations as the pair

$$u_x = xv_y \text{ and } u_y = v_x \tag{4.5.2}$$

we find $k^+(x, y) = -x^{-1/2}$, $k^-(x, y) = x^{-1/2}$, $\gamma^+ = \gamma^- = 0$. Thus on the curve $y = \frac{2}{3}x^{3/2} + c$ there is a differential relation $x^{-1/2}du - dv = 0$, while on the curve $y = c - \frac{2}{3}x^{3/2}$, $x^{-1/2}du + dv = 0$.

(*e*) Finally if we write this equation as

$$(x^2 - 1)u_x + xy(u_y + v_x) + (y^2 - 1)v_y = -2xu - 2yv, \; u_y = v_x, \tag{4.5.3}$$

then
$$\{xy - \sqrt{(x^2 + y^2 - 1)}\}k^+(x, y) = x^2 - 1,$$
$$\{xy + \sqrt{(x^2 + y^2 - 1)}\}k^-(x, y) = x^2 - 1,$$
$$\{y + x\sqrt{(x^2 + y^2 - 1)}\}\gamma^+ = -2(xu + yv)k^+(x, y),$$
and
$$\{y - x\sqrt{(x^2 + y^2 - 1)}\}\gamma^- = -2(xu + yv)k^-(x, y)$$
and we find that on the line $(x^2 - 1)dy/dx = xy + \sqrt{(x^2 + y^2 - 1)}$

$$(x^2 - 1)\frac{du}{ds} + \{xy - \sqrt{(x^2 + y^2 - 1)}\}\frac{dv}{ds} - 2\,\frac{y - x\sqrt{(x^2 + y^2 - 1)}}{x^2 + y^2}(xu + yv) = 0,$$

with an equivalent result on the other characteristic, the sign of the square root being changed.

4.6 Reduction to canonical form

The final topic that we shall consider in this chapter is the reduction of a linear equation like (4.1.1) to canonical form. Let us suppose first that the equation is hyperbolic so that the equation

$$A(x, y)\zeta^2 - 2B(x, y)\zeta + C(x, y) = 0$$

has the two real roots

$$A(x, y)\zeta^+ = B(x, y) + H(x, y) \text{ and } A(x, y)\zeta^- = B(x, y) - H(x, y)$$

where we have put

$$\{B(x, y)\}^2 = \{H(x, y)\}^2 + A(x, y)\, C(x, y).$$

We choose a function $\xi(x, y)$ to be a solution of the first-order partial equation

$$A(x, y)\frac{\partial \xi}{\partial x} + \{B(x, y) + H(x, y)\}\frac{\partial \xi}{\partial y} = 0. \tag{4.6.1}$$

This means that the curves $\xi(x, y) =$ const. will generate the family of characteristics which are integral curves of $dy/dx = \zeta^+$. We can in a similar way find a function $\eta(x, y)$ constant on each of the other family of characteristics by choosing it to satisfy

$$A(x, y)\frac{\partial \eta}{\partial x} + \{B(x, y) - H(x, y)\}\frac{\partial \eta}{\partial y} = 0. \tag{4.6.2}$$

We take ξ and η as new curvilinear coordinates and transform the equation accordingly. Under a general non-degenerate transformation of this type, (4.1.1) will become

$$\begin{aligned} A'(\xi, \eta)\Phi_{\xi\xi} + 2B'(\xi, \eta)\Phi_{\xi\eta} + C'(\xi, \eta)\Phi_{\eta\eta} + E'(\xi, \eta)\Phi_{\xi} \\ + F'(\xi, \eta)\Phi_{\eta} + G'(\xi, \eta)\Phi = D'(\xi, \eta), \end{aligned} \tag{4.6.3}$$

where

$$A'(\xi, \eta) = A\left(\frac{\partial \xi}{\partial x}\right)^2 + 2B\frac{\partial \xi}{\partial x}\frac{\partial \xi}{\partial y} + C\left(\frac{\partial \xi}{\partial y}\right)^2,$$

$$B'(\xi, \eta) = A\frac{\partial \xi}{\partial x}\frac{\partial \eta}{\partial x} + B\left(\frac{\partial \xi}{\partial x}\frac{\partial \eta}{\partial y} + \frac{\partial \xi}{\partial y}\frac{\partial \eta}{\partial x}\right) + C\frac{\partial \xi}{\partial y}\frac{\partial \eta}{\partial y},$$

$$C'(\xi, \eta) = A\left(\frac{\partial \eta}{\partial x}\right)^2 + 2B\frac{\partial \eta}{\partial x}\frac{\partial \eta}{\partial y} + C\left(\frac{\partial \eta}{\partial y}\right)^2,$$

$$E'(\xi, \eta) = A\frac{\partial^2 \xi}{\partial x^2} + 2B\frac{\partial^2 \xi}{\partial x\, \partial y} + C\frac{\partial^2 \xi}{\partial y^2} + E\frac{\partial \xi}{\partial x} + F\frac{\partial \xi}{\partial y},$$

$$F'(\xi, \eta) = A\frac{\partial^2 \eta}{\partial x^2} + 2B\frac{\partial^2 \eta}{\partial x\, \partial y} + C\frac{\partial^2 \eta}{\partial y^2} + E\frac{\partial \eta}{\partial x} + F\frac{\partial \eta}{\partial y},$$

$$G'(\xi, \eta) = G(x, y) \text{ and } D'(\xi, \eta) = D(x, y).$$

If then ξ is chosen to satisfy (4.6.1) it will cause A' to vanish, and if η is chosen to satisfy (4.6.2) it will cause C' to vanish. The first canonical form of a linear second-order equation of hyperbolic type is therefore just

$$2B'(\xi, \eta)\Phi_{\xi\eta} + E'(\xi, \eta)\Phi_\xi + F'(\xi, \eta)\Phi_\eta + G'(\xi, \eta)\Phi = D'(\xi, \eta). \tag{4.6.4}$$

If we put $2\xi = X + Y$, $2\eta = X - Y$, (4.6.4) is replaced by

$$2B''(X, Y)\{\Phi_{XX} - \Phi_{YY}\} + E''(X, Y)\Phi_X + F''(X, Y)\Phi_Y + G''(X, Y)\Phi = D''(X, Y) \tag{4.6.5}$$

which is the second canonical form, where B'', G'' and D'' are obtained directly from B', G' and D' by substituting the new variables and

$$E'' = E' + F',\ F'' = E' - F'.$$

If (4.1.1) had been elliptic, the corresponding roots of the fundamental quadratic could have been written

$$A\zeta^+ = B + iH \text{ and } A\zeta^- = B - iH,$$

wherein we have now put $\{B\}^2 + \{H\}^2 = AC$. If, corresponding to (4.6.1) and (4.6.2), we choose ξ to be a solution of

$$A\frac{\partial \xi}{\partial x} + \{B + iH)\frac{\partial \xi}{\partial y} = 0, \tag{4.6.6}$$

and η to be a solution of

$$A\frac{\partial \eta}{\partial x} + \{B - iH\}\frac{\partial \eta}{\partial y} = 0, \tag{4.6.7}$$

ξ and η can be paired as complex conjugates, and if we use them as coordinates we shall derive an equation like (4.6.4), but in which the coefficients B', G' and D' will be real, while those of E' and F' will be complex conjugate. The final transformation then

$$\xi = \tfrac{1}{2}(X + iY) \text{ and } \eta = \tfrac{1}{2}(X - iY)$$

will produce a form like (4.6.5), namely

$$2B''\{\Phi_{XX} + \Phi_{YY}\} + E''\Phi_X + F''\Phi_Y + G''\Phi = D'', \tag{4.6.8}$$

in which all the coefficients are real. This then is the canonical form for the elliptic equation.

Finally if (4.1.1) had been parabolic we would have had

$$(B)^2 = AC,$$

and so A and C would have had the same sign. For convenience this will be taken as positive, and we shall also assume that B is positive. We may then choose ξ to satisfy

$$\sqrt{A}\frac{\partial \xi}{\partial x} + \sqrt{C}\frac{\partial \xi}{\partial y} = 0 \tag{4.6.9}$$

which will mean that $A' = 0$. Also

$$B' = \left(\sqrt{A}\frac{\partial \xi}{\partial x} + \sqrt{C}\frac{\partial \xi}{\partial y}\right)\left(\sqrt{A}\frac{\partial \eta}{\partial x} + \sqrt{C}\frac{\partial \eta}{\partial y}\right) = 0,$$

and so, if we may also select η to satisfy

$$A\frac{\partial^2 \eta}{\partial x^2} + 2B\frac{\partial^2 \eta}{\partial x\,\partial y} + C\frac{\partial^2 \eta}{\partial y^2} + E\frac{\partial \eta}{\partial x} + F\frac{\partial \eta}{\partial y} = 0,$$

the equation reduces to

$$C'(\xi, \eta)\Phi_{\eta\eta} + E'(\xi, \eta)\Phi_{\xi} + G(\xi, \eta)\Phi = D'(\xi, \eta), \qquad (4.6.10)$$

which is the canonical form of the parabolic equation.

4.7 The canonical form of the examples

We shall again use the last two examples of Section 4.3 to illustrate the reduction to canonical form. Taking first the Tricomi equation, then in the half-plane $x > 0$ we define new coordinates by

$$\xi = \tfrac{1}{2}(y + \tfrac{2}{3}x^{3/2}) \text{ and } \eta = \tfrac{1}{2}(\tfrac{2}{3}x^{3/2} - y).$$

In terms of these new and independent variables the equation becomes

$$6(\xi + \eta)\Phi_{\xi\eta} + \Phi_{\xi} + \Phi_{\eta} = 0,$$

a Darboux equation. If we put $X = \tfrac{2}{3}x^{3/2}$, $Y = y$,

$$3X(\Phi_{XX} - \Phi_{YY}) + \Phi_X = 0.$$

In the half-plane $x < 0$ we take as new coordinates $X = \tfrac{2}{3}(-x)^{3/2}$, $Y = y$, when the equation reduces to

$$3X(\Phi_{XX} + \Phi_{YY}) + \Phi_X = 0.$$

In the cone field problem, the hyperbolic region is $r > 1$, and in this we take as new coordinates

$$\xi = \frac{1}{2}\left\{\cos^{-1}\left(\frac{1}{r}\right) + \theta\right\},\ \eta = \frac{1}{2}\left\{\cos^{-1}\left(\frac{1}{r}\right) - \theta\right\}$$

which reduces the equation simply to

$$\Phi_{\xi\eta} = 0.$$

In consequence if $X = \cos^{-1}(1/r)$, $Y = \theta$, we get the ordinary wave equation

$$\Phi_{XX} - \Phi_{YY} = 0.$$

In a similar manner, inside the unit circle we can take as new coordinates $X = \cosh^{-1}(1/r)$, $Y = \theta$, and the equation transforms into the Laplace equation

$$\Phi_{XX} + \Phi_{YY} = 0.$$

4.8 Classification of quasi-linear equations

All the equations we have examined in this chapter have been either linear equations or semi-linear equations in two independent variables. The results may be extended to quasi-linear equations, in an analogous manner to that by which our results for semi-linear first-order equations were extended to cover quasi-linear ones. We find, as then, that the characteristics do themselves depend on the solution of the partial equation, and are not given in advance as two sets of curves covering the (x, y) plane, as in the examples we have considered above. The ordinary differential equations determining the characteristics, and the total differential equations on the characteristics, must then be treated together as simultaneous differential equations.

The best-known examples of quasi-linear partial differential equations are those of classical hydrodynamics, which we shall discuss in a later chapter.

In the next chapter we shall consider linear equations in more than two independent variables.

EXERCISES

4.1. Show that the equation

$$y^2\frac{\partial^2\Phi}{\partial x^2} - x^2\frac{\partial^2\Phi}{\partial y^2} = 0$$

is hyperbolic in every domain away from the axes $x = 0, y = 0$, and that the two families of characteristics are the circles $x^2 + y^2 = c_1$ and the hyperbolae $x^2 - y^2 = c_2$.

Show further that it has the canonical form

$$2(\xi^2 - \eta^2)\frac{\partial^2\Phi}{\partial\xi\,\partial\eta} + \eta\frac{\partial\Phi}{\partial\xi} + \xi\frac{\partial\Phi}{\partial\eta} = 0.$$

4.2. Show that the equation

$$(1 - x^2)\frac{\partial^2\Phi}{\partial x^2} - \frac{\partial^2\Phi}{\partial y^2} = 0$$

is hyperbolic in the region $-1 < x < 1$, $-\infty < y < \infty$, and find the characteristics. Derive also the canonical form

$$2\frac{\partial^2\Phi}{\partial\xi\,\partial\eta} = \tan(\xi - \eta)\left\{\frac{\partial\Phi}{\partial\xi} - \frac{\partial\Phi}{\partial\eta}\right\}.$$

4.3. Show that

$$\frac{\partial^2\Phi}{\partial x^2} + xy\frac{\partial^2\Phi}{\partial y^2} = 0$$

is hyperbolic in two quadrants, and elliptic in the other two quadrants of the plane. Find the characteristics in the quadrants in which the equation is hyperbolic, and derive the canonical forms

$$\frac{\partial^2\Phi}{\partial\xi^2} \pm \frac{\partial^2\Phi}{\partial\eta^2} + \frac{1}{3\xi}\frac{\partial\Phi}{\partial\xi} \mp \frac{1}{\eta}\frac{\partial\Phi}{\partial\eta} = 0.$$

4.4. Show that the equation

$$y^2\frac{\partial^2\Phi}{\partial x^2} + 2xy\frac{\partial^2\Phi}{\partial x\,\partial y} + x^2\frac{\partial^2\Phi}{\partial y^2} - x\frac{\partial\Phi}{\partial x} - y\frac{\partial\Phi}{\partial y} = 0$$

is parabolic, and reduce it to canonical form.

ANSWERS

4.2. $x = \pm \sin\{y - c\}$.

4.3. The equation is hyperbolic in the 2nd and 4th quadrants with characteristics

$$2\sqrt{|y|} = C \pm \tfrac{2}{3}|x|^{3/2}.$$

4.4. $(\eta^2 - \xi^2)\dfrac{\partial^2\Phi}{\partial\eta^2} - \xi\dfrac{\partial\Phi}{\partial\xi} = 0.$

CHAPTER 5

Linear Partial Differential Equations of Second Order in more than Two Independent Variables

5.1 The Cauchy problem

We shall take the independent variables as $x_1, x_2, \ldots, x_n$ and assume that they form the components of an n-dimensional vector $\mathbf{x}$. If we use the summation convention, a general second-order linear equation has the form

$$a_{ij}(\mathbf{x})\,\Phi_{ij} + b_i(\mathbf{x})\,\Phi_i + c(\mathbf{x})\,\Phi(\mathbf{x}) = f(\mathbf{x}), \tag{5.1.1}$$

where

$$\Phi_i = \frac{\partial \Phi}{\partial x_i} \text{ and } \Phi_{ij} = \frac{\partial^2 \Phi}{\partial x_i\,\partial x_j}.$$

The corresponding Cauchy problem would be to find a solution of this equation (5.1.1) which took the values of a known function on a segment of a given $(n-1)$-dimensional hypersurface Σ, and whose first partial derivative in the direction normal to the hypersurface was also known at each point of the segment. The Cauchy-Kowalewski theorem would of course, at most, only assert the existence of the solution in a domain immediately adjacent to the surface.

In this case we cannot treat the equation (5.1.1) as the equivalent of two equations of first order. It is easy to see that if the number of independent variables is greater than two, one equation of second order can only be replaced by a set of as many equations as there are independent variables. Similarly a pair of simultaneous first-order equations can only be reduced to a single equation of order greater than two.

5.2 The solution of the Cauchy problem

We must make some assumptions about the regularity of the hypersurface Σ. We will suppose that the segment we are interested in is an $(n-1)$-dimensional neighbourhood of a point P on the surface, and that if the surface is defined by a relation of the form

$$G(\mathbf{x}) = 0, \tag{5.2.1}$$

this may be used to find all the $(n-1)$ first partial derivatives of one of the x_i with respect to the other $(n-1)$ independent variables in this neighbourhood of P. Furthermore we assume these partial derivatives are themselves continuous functions of their arguments in the neighbourhood.

We conclude that we may introduce a set of new curvilinear coordinates in the neighbourhood $\boldsymbol{\xi} = (\xi_1, \xi_2, \ldots \xi_n)$, with the properties:

(*a*) P is the point $\boldsymbol{\xi} = 0$,
(*b*) Σ is (locally) the surface $\xi_n = 0$,
(*c*) At P the direction ξ_n increasing is normal to Σ.

In terms of the new coordinates the equation becomes

$$A_{ij}(\boldsymbol{\xi})\frac{\partial^2\Phi}{\partial\xi_i\,\partial\xi_j} + B_i(\boldsymbol{\xi})\frac{\partial\Phi}{\partial\xi_i} + C(\boldsymbol{\xi})\Phi = F(\boldsymbol{\xi}), \tag{5.2.2}$$

where

$$A_{ij}(\boldsymbol{\xi}) = a_{kl}\frac{\partial\xi_i}{\partial x_k}\frac{\partial\xi_j}{\partial x_l},$$

$$B_i(\boldsymbol{\xi}) = b_j\frac{\partial\xi_i}{\partial x_j} + a_{kl}\frac{\partial^2\xi_i}{\partial x_k\,\partial x_l}$$

$$C(\boldsymbol{\xi}) = c(\mathbf{x}) \text{ and } F(\boldsymbol{\xi}) = f(\mathbf{x}).$$

If then we may assume that Φ and its normal derivative are known on Σ near P, this is equivalent to knowing all the partial derivatives which exist and belong to one of the two sets:

(*a*) All mixed partial derivatives with respect to $\xi_1, \xi_2, \ldots, \xi_{n-1}$
(*b*) $\partial/\partial\xi_n$ of all such mixed derivatives with respect to $\xi_1, \xi_2, \ldots, \xi_{n-1}$.

Hence (5.2.2) is an equation in which all the terms may be regarded as known on the surface, with the exception of

$$\frac{\partial^2\Phi}{\partial{\xi_n}^2}.$$

The condition that this can be found is therefore that its coefficient does not vanish. If, as in the proof of the Cauchy-Kowalewski theorem, we are able to differentiate (5.2.2) successively, the same condition will allow us to find the higher derivatives of the form

$$\frac{\partial^m \Phi}{\partial \xi_n{}^m},$$

and so to continue the solution of the equation away from the hypersurface.

This condition is then

$$A_{nn}(\xi) = a_{kl}\frac{\partial \xi_n}{\partial x_k}\frac{\partial \xi_n}{\partial x_l} \neq 0. \tag{5.2.3}$$

5.3 Classification of the equations

The set of partial derivatives $(\partial \xi_n / \partial x_i)$ can be regarded as the components of a vector which is clearly parallel to the unit vector $\mathbf{n}$ normal to the surface Σ. We therefore find that, associated with the point P, is a family of unit vectors $\boldsymbol{\eta} = (\eta_j)$ which satisfies the equation

$$a_{ij}\eta_i\eta_j = 0. \tag{5.3.1}$$

If we draw through P all straight lines with directions parallel to one of the vectors $\boldsymbol{\eta}$ which satisfies (5.3.1), they will be the family of generators of a quadric cone with P as its vertex. This we call the normal cone.

If (5.3.1) is either a positive definite, or negative definite quadratic form, then obviously the cone is purely imaginary, and we then say that the original partial differential equation is elliptic.

In all other cases we might expect the cone to have at least some real sheets. By a well-known theorem of linear algebra [3], (5.3.1) can be reduced by a real linear transformation to a quadratic form in which no cross-product terms appear. If this canonical form contains less than the full n terms, we say that the equation is parabolic. If it contains the full number of terms, and the coefficients are of differing signs, we say that it is hyperbolic. If, finally, the coefficients are all of one sign, with one exception only, we say that the equation is of normal hyperbolic type.

It is a matter of observation that most of the equations which appear in mathematical physics are elliptic, of normal hyperbolic type, or parabolic of a special sort in which the canonical form, although in a reduced number of variables, has all its terms of one sign only, and so is semi-definite.

We may in fact use the transformation of coordinates, not only to remove the cross-product terms, but also to reduce each of the other coefficients to ± 1. It is again well known from algebraic theory that the number of positive coefficients and the number of negative coefficients appear as invariants of the transformation. The result of the transformation, therefore, will be to reduce the original to a canonical form comparable with (4.6.5), (4.6.8) or (4.6.10).

5.4 Characteristic surfaces and bicharacteristics

Just as in the previous chapter we saw that the Cauchy problem was, in general, only properly posed, that is representing a soluble problem with a unique solution, if the equation was hyperbolic, so in this case it will be properly posed only for the equation of normal hyperbolic form, since this has a normal cone which is everywhere real.

Corresponding to each generator of the normal cone, we can draw through P a plane whose normal is parallel to the generator. Such planes will themselves envelop another cone which is called the characteristic cone at P. Any surface in space, with the property that at each point its tangent plane also touches the characteristic cone associated with the point, is said to be a characteristic surface.

If a twisted curve is drawn in space so that at each point its tangent vector is directed along a generator of the characteristic cone there, we say the curve is a bi-characteristic. All bi-characteristics through a point P lie in a surface through P, sometimes called the characteristic conoid at P. Obviously at P the characteristic cone and the characteristic conoid are asymptotically coincident. If the original equation has constant coefficients the characteristic conoid and the cone will coincide everywhere. It must be stated unequivocally that this use of the word conoid does not correspond to the usual geometric definition.

We see that the following statements must be true, just as we noticed in the previous chapter for their two-dimensional analogues.

(*a*) The values at P of a function and its derivative normal to a given surface Σ through P will only affect values of a solution of the equation, taking the given values on Σ, at points which lie within or on the characteristic conoid through P.

(*b*) Conversely, the value taken by a function which is a solution of the equation, at a point P not on Σ, is determined only by the boundary conditions at points within the intercept made on Σ by the conoid through P.

(*c*) If the surface Σ on which the boundary conditions are given is everywhere characteristic, the solution cannot be continued beyond it.

(*d*) Any discontinuity in the solution must therefore take place across a characteristic surface.

This last statement is equivalent to equating characteristic surfaces with the more intuitive physical idea of a wave front on which a disturbance will be propagated.

5.5 The adjoint equation

With a linear equation of second order we can associate a second equation, its adjoint equation. Let V be a volume of n-dimensional space bounded by a closed $(n-1)$-dimensional surface Σ, which possesses at each point, except possibly at a set of points of zero measure, a unique outward normal vector.

If we use the n-dimensional Gauss theorem (see Appendix 1), we see that, if all the coefficients of the equation are doubly differentiable, and $\psi(\mathbf{x})$ is also doubly differentiable:

$$\int_V \psi(\mathbf{x})\{a_{ij}(\mathbf{x})\,\Phi_{ij} + b_i(\mathbf{x})\,\Phi_i + c(\mathbf{x})\,\Phi(\mathbf{x})\}\,d\tau$$

$$= \int_\Sigma \psi(\mathbf{x})\{a_{ij}(\mathbf{x})\,\Phi_j + b_i(\mathbf{x})\,\Phi(\mathbf{x})\}n_i\,dS$$

$$-\int_V \left[\Phi_j \frac{\partial}{\partial x_i}\{a_{ij}(\mathbf{x})\,\psi(\mathbf{x})\} - \Phi(\mathbf{x})\frac{\partial}{\partial x_i}\{b_i(\mathbf{x})\,\psi(\mathbf{x})\} - c(\mathbf{x})\,\Phi(\mathbf{x})\,\psi(\mathbf{x})\right] d\tau$$

$$= \int_\Sigma \left[\psi(\mathbf{x})\{a_{ij}(\mathbf{x})\,\Phi_j + b_i(\mathbf{x})\,\Phi(\mathbf{x})\}n_i - \Phi(\mathbf{x})\,n_j\frac{\partial}{\partial x_i}\{a_{ij}(\mathbf{x})\psi(\mathbf{x})\}\right] dS$$

$$+\int_V \Phi(\mathbf{x})\left[\frac{\partial^2}{\partial x_i\,\partial x_j}\{a_{ij}(\mathbf{x})\,\psi(\mathbf{x})\} - \frac{\partial}{\partial x_i}\{b_i(\mathbf{x})\,\psi(\mathbf{x})\} + c(\mathbf{x})\,\psi(\mathbf{x})\right] d\tau.$$

If we put

$$L(\Phi) = a_{ij}(\mathbf{x})\,\Phi_{ij} + b_i(\mathbf{x})\,\Phi_i + c(\mathbf{x})\,\Phi,$$

and

$$M(\psi) = \frac{\partial^2}{\partial x_i\,\partial x_j}\{a_{ij}(\mathbf{x})\,\psi(\mathbf{x})\} - \frac{\partial}{\partial x_i}\{b_i(\mathbf{x})\,\psi(\mathbf{x})\} + c(\mathbf{x})\,\psi(\mathbf{x})$$

then

$$\int_V \{\psi(\mathbf{x})\,L(\Phi) - \Phi(\mathbf{x})\,M(\psi)\}\,d\tau = \int_\Sigma \Big[a_{ij}(\mathbf{x})\,\{n_i\psi(\mathbf{x})\,\Phi_j - n_j\Phi(\mathbf{x})\,\psi_i\}$$
$$+\left\{n_i b_i(\mathbf{x}) - n_j\frac{\partial a_{ij}}{\partial x_i}\right\}\psi(\mathbf{x})\,\Phi(\mathbf{x})\Big]\,dS. \qquad (5.5.1)$$

If as is usual $a_{ij}(\mathbf{x}) = a_{ji}(\mathbf{x})$, so that the coefficient matrix is symmetric, we may simplify the right-hand side.

Introduce the vector $\mathbf{\Lambda}$ with components $\Lambda_j = a_{ij}n_i$, and let Λ be the modulus of the vector, $\boldsymbol{\nu}$ a unit vector in the same direction as $\mathbf{\Lambda}$. Then the right-hand side is

$$\int_\Sigma \left[a_{ij}(\mathbf{x})\,\{\psi(\mathbf{x})\,\Phi_j - \Phi(\mathbf{x})\,\psi_j\} + \left\{b_i(\mathbf{x}) - \frac{\partial a_{ij}}{\partial x_j}\right\}\psi(\mathbf{x})\,\Phi(\mathbf{x})\right] n_i\,dS,$$

$$= \int_\Sigma \left[\left(\psi(\mathbf{x})\frac{\partial \Phi}{\partial \nu} - \Phi(\mathbf{x})\frac{\partial \psi}{\partial \nu}\right)\Lambda + \left\{b_i(\mathbf{x}) - \frac{\partial a_{ij}}{\partial x_j}\right\}\psi(\mathbf{x})\,\Phi(\mathbf{x})\,n_i\right] dS.$$
$$(5.5.2)$$

The equation

$$M\{\psi) = 0$$

is said to be the adjoint equation to (5.1.1). If this coincides with (5.1.1) the equation is said to be self-adjoint. The condition that this is true is that

$$2\frac{\partial a_{ij}}{\partial x_j} - b_i(\mathbf{x}) = b_i(\mathbf{x})$$

and

$$\frac{\partial^2 a_{ij}}{\partial x_i\, \partial x_j} - \frac{\partial b_i}{\partial x_i} + c(\mathbf{x}) = c(\mathbf{x}).$$

Both these are clearly satisfied if

$$b_i = \frac{\partial a_{ij}}{\partial x_j}, \tag{5.5.3}$$

and when this holds (5.1.1) can be written

$$\frac{\partial}{\partial x_j}\{a_{ij}(\mathbf{x})\, \Phi_i\} + c(\mathbf{x})\, \Phi = f(\mathbf{x}), \tag{5.5.4}$$

which is therefore the standard form of a self-adjoint second-order linear equation.

The importance of the adjoint equation will appear in a subsequent chapter when we consider the construction of Green's functions. A large number of the equations in which the mathematical physicist is interested are self-adjoint.

5.6 The conormal

The vector

$$\boldsymbol{\nu} = (\nu_j) = A^{-1}(a_{ij} n_i)$$

that we introduced in the previous section is a unit vector whose direction is said to be that of the conormal. Its direction can be related geometrically to the direction of the normal $\mathbf{n}$.

If we consider the quadric surface

$$a_{ij}\xi_i\xi_j = \varepsilon^2, \tag{5.6.1}$$

then it has a tangent plane at the point $(\boldsymbol{\xi}^0)$

$$a_{ij}\xi_i{}^0\xi_j = \varepsilon^2. \tag{5.6.2}$$

Hence if we draw a vector from the centre of the quadric in the direction of $\mathbf{n}$, then $\boldsymbol{\nu}$ will be in the direction of the normal to the diametrical plane of the quadric conjugate to $\mathbf{n}$.

In the particular case of an equation of normal hyperbolic type, (5.6.2) still represents a non-degenerate quadric surface even when $\varepsilon = 0$. It will then follow that the conormal will be directed along a generator of the characteristic cone, and so will be directed along the tangent to a bicharacteristic at the point.

Since also

$$\Lambda v_j n_j = a_{ij} n_i n_j, \tag{5.6.3}$$

if the surface Σ is characteristic

$$v_j n_j = 0, \tag{5.6.4}$$

and the conormal is a tangent to the surface.

Exercises

5.1. State whether the following equations are elliptic, of normal hyperbolic type or parabolic (suffixes denote differentiations):

(*a*) $3\Phi_{xx} + 2\Phi_{yy} + 3\Phi_{zz} + 2\Phi_{xz} = 0$
(*b*) $4\Phi_{xx} + 4\Phi_{yy} + 2\Phi_{zz} + 6\Phi_{xy} + 2\Phi_{yz} - 2\Phi_{xz} = 0$
(*c*) $2\Phi_{xx} + 4\Phi_{yy} + 8\Phi_{zz} + 6\Phi_{xy} + 10\Phi_{yz} + 6\Phi_{xz} = 0$
(*d*) $5\Phi_{xx} + 8\Phi_{yy} + 3\Phi_{zz} + 6\Phi_{xy} + 2\Phi_{yz} + 4\Phi_{xz} = 0$
(*e*) $3\Phi_{xx} + 2\Phi_{yy} + 2\Phi_{zz} - 4\Phi_{xy} + 2\Phi_{yz} - 4\Phi_{xz} = 0.$

5.2. State whether the following are elliptic, of normal hyperbolic type, hyperbolic but not of normal type, or parabolic:

(*a*) $\Phi_{xx} + \Phi_{yy} + \Phi_{zz} + \Phi_{ww} - \Phi_{xy} - \Phi_{yz} - \Phi_{zw} - \Phi_{xw} = 0$
(*b*) $\Phi_{xx} + 13\Phi_{yy} + 13\Phi_{zz} + \Phi_{ww} + 4\Phi_{xy} + 24\Phi_{yz} + 4\Phi_{zw} = 0$
(*c*) $4\Phi_{xx} + \Phi_{zz} + 8\Phi_{ww} + 4\Phi_{xy} + 2\Phi_{yz} + 12\Phi_{xw} = 0.$

5.3. Show that the equation

$$(y^2 - z^2)\Phi_{xx} + (z^2 - x^2)\Phi_{yy} + (x^2 - y^2)\Phi_{zz} = 0$$

is of normal hyperbolic type almost everywhere. Show further that for all values of C: (*a*)$x \pm y \pm z = C$, and (*b*) $x^2 \pm y^2 \pm z^2 = C$ are characteristic surfaces.

5.4. Show that the equation

$$yz\Phi_{xx} + zx\Phi_{yy} + xy\Phi_{zz} = 0$$

is elliptic in the two octants where $x > 0$, $y > 0$, $z > 0$ and where $x < 0$, $y < 0$, $z < 0$, and is hyperbolic almost everywhere else. In the octant $x < 0$, $y > 0$, $z > 0$ show that

$$\sqrt{2}(-x)^{3/2} \pm y^{3/2} \pm z^{3/2} = C$$

are characteristic surfaces.

5.5. Find the adjoint equations of

(*a*) $x^2\Phi_{xx} + y^2\Phi_{yy} + z^2\Phi_{zz} + 2xy\Phi_{xy} + 2yz\Phi_{yz} + 2xz\Phi_{zx} + 2x\Phi_x + 2y\Phi_y + 2z\Phi_z = 0$

(*b*) $(x^2 + y^2)\Phi_{xx} + (y^2 + z^2)\Phi_{zz} + x^3y^3z^3\Phi y = 0.$

5.6. Show that

$$x^3\Phi_{xx} + (y^2 + yz)\Phi_{yy} + (x + y)^2\Phi_{zz} + 3x^2\Phi_x + (2y + z)\Phi_y = 0$$

is self-adjoint.

ANSWERS

5.1. (*a*) Elliptic, (*b*) parabolic, (*c*) hyperbolic, (*d*) hyperbolic, (*e*) elliptic.

5.2. (*a*) Elliptic, (*b*) hyperbolic (non-normal), (*c*) parabolic.

5.5. (*a*) $x^2\Phi_{xx} + y^2\Phi_{yy} + z^2\Phi_{zz} + 2xy\Phi_{xy} + 2yz\Phi_{yz} + 2xz\Phi_{xz} + 6x\Phi_x + 6y\Phi_y + 6z\Phi_z + 6\Phi = 0$

(*b*) $(x^2 + y^2)\Phi_{xx} + (y^2 + z^2)\Phi_{zz} + 2x\Phi_x - x^3y^3z^3\Phi_y + 2z\Phi_z + (4 - 3x^3y^3z^3)\Phi = 0.$

CHAPTER 6

The Laplace Equation

6.1 The linear properties

The form of the Laplace equation in a space of n dimensions is

$$\nabla_n^2 \Phi = \frac{\partial^2 \Phi}{\partial x_i \, \partial x_i} = 0, \tag{6.1.1}$$

which can be obtained from the canonical form of the general elliptic linear equation of second order by neglecting all but the principal terms. In this sense therefore the equation represents the typical elliptic equation. Any solution of (6.1.1) we will call a harmonic function in n-dimensional space. In spaces of two and three dimensions the Laplace equation is familiar in mathematical physics as being the governing equation of certain problems in hydrodynamics, electrostatics and magnetostatics. For this reason we shall confine the investigation of particular forms of its solution, later in this chapter, to these two cases. In these cases the solution represents a potential function.

The equation is linear and homogeneous, and so possesses the two properties always associated with linear systems.

Property A: If $\Phi_1(\mathbf{x})$ and $\Phi_2(\mathbf{x})$ are both solutions of equation (6.1.1), then so also is

$$a\Phi_1(\mathbf{x}) + b\Phi_2(\mathbf{x}),$$

for all constants a and b.

This is often expressed by the statement that solutions of (6.1.1) are additive.

Property B: If $\Psi(\mathbf{x})$ is a particular solution of the inhomogeneous equation

$$\nabla_n^2 \Psi = -\Omega_n \rho(\mathbf{x}), \tag{6.1.2}$$

where $\rho(\mathbf{x})$ is a known function and Ω_n is the area of the surface of the unit sphere in n dimensions (see Appendix 2), then all solutions of (6.1.2) have the form

$$\Psi(\mathbf{x}) + \Phi(\mathbf{x}),$$

where $\Phi(\mathbf{x})$ is a solution of (6.1.1).

The inhomogeneous equation is usually designated the Poisson equation. The term on its right-hand side just represents some known function of the vector $\mathbf{x}$; the only reason for choosing that particular form is that this is the form in which it commonly appears in mathematical physics.

6.2 The averaging properties

The Laplacian operator has a well-known averaging property. This can be seen by writing in the n-dimensional Gauss theorem (see Appendix 1)

$$X_i(\mathbf{x}) = \frac{\partial \Phi}{\partial x_i}, \tag{6.2.1}$$

and applying the theorem to the n-dimensional sphere of radius r

$$x_i x_i = r^2. \tag{6.2.2}$$

Since the normal derivative in the outward direction on the sphere is just $\partial\Phi/\partial r$, the statement of the theorem reduces to

$$\int_{x_i x_i < r^2} \{\nabla_n{}^2 \Phi(\mathbf{x})\}\, d\tau = \int_{x_i x_i = r^2} \frac{\partial \Phi}{\partial r}\, dS. \tag{6.2.3}$$

Now suppose that, in some domain including the origin, $\Phi(\mathbf{x})$ satisfies (6.1.1), then for every sphere like (6.2.2) that can be drawn in the domain we must have

$$0 = \int_{x_i x_i = r^2} \frac{\partial \Phi}{\partial r}\, dS = r^{n-1} \int_{x_i x_i = 1} \frac{\partial \Phi}{\partial r}\, d\Omega_n, \tag{6.2.4}$$

if Ω_n is again the surface area of the unit sphere in n dimensions (see Appendix 2). If, however, differentiation under the integral sign is permitted (and this will apply provided we remain in the interior of the domain in which (6.1.1) is satisfied), the right-hand side of (6.2.4) is:

$$r^{n-1} \frac{\partial}{\partial r} \int_{x_i x_i = 1} \Phi(\mathbf{x})\, d\Omega_n = \Omega_n r^{n-1} \frac{\partial}{\partial r} \left\{ \frac{1}{\Omega_n} \int_{x_i x_i = 1} \Phi(\mathbf{x})\, d\Omega_n \right\}$$

$$= \Omega_n r^{n-1} \frac{\partial}{\partial r} \left\{ \frac{1}{\Omega_n r^{n-1}} \int_{x_i x_i = r^2} \Phi(\mathbf{x})\, dS \right\}.$$

The quantity in brackets, which we may denote by $M_r\{\Phi\}$, is the average value of $\Phi(\mathbf{x})$ taken over the surface of a sphere of radius r. Hence (6.2.4) can be written

$$\frac{\partial}{\partial r} M_r\{\Phi\} = 0, \tag{6.2.5}$$

or alternatively $M_r\{\Phi\}$ is a constant. In particular the value of the constant can be taken as $M_0\{\Phi\}$, or, which is the same thing, $\Phi(\mathbf{0})$. Hence if $\Phi(\mathbf{x})$ satisfies the Laplace equation, its average value, taken over the surface of any sphere, is the value at the centre of the sphere. This means either that $\Phi(\mathbf{x})$ is a constant in and on the sphere, or that on the surface it takes values both greater and less than those at the centre. Since this is true for every such sphere, $\Phi(\mathbf{x})$ can neither take a maximum nor a minimum value at points inside a domain in which it satisfies (6.1.1).

6.3 Uniqueness theorem

Before we start on the examination of particular solutions in two or three dimensions, we wish to establish some general results. For this purpose we use Green's theorem, which may be deduced from the Gauss theorem (see Appendix 1). If we substitute into the theorem

$$X_i(\mathbf{x}) = \Psi(\mathbf{x})\frac{\partial \Phi}{\partial x_i} = \Psi(x)\,\Phi_i$$

we deduce that

$$\int_V \{\Psi(\mathbf{x})\nabla_n{}^2\Phi + \Psi_i\Phi_i\}\,d\tau = \int_\Sigma \Psi(\mathbf{x})\,\Phi_i n_i\,dS, \qquad (6.3.1)$$

where n_i as usual are the components of the unit outward normal to the bounding surface Σ. We can obviously produce a similar equation with the roles of $\Psi(\mathbf{x})$ and $\Phi(\mathbf{x})$ reversed, so that by subtraction we deduce the identity of Green:

$$\int_V \{\Psi\nabla_n{}^2\Phi - \Phi\nabla_n{}^2\Psi\}\,d\tau = \int_\Sigma \left\{\Psi(\mathbf{x})\frac{\partial \Phi}{\partial n} - \Phi(\mathbf{x})\frac{\partial \Psi}{\partial n}\right\} dS, \qquad (6.3.2)$$

where $\partial/\partial n$ signifies differentiation along the outward normal to Σ.

The first use that we can make of this is to deduce a uniqueness theorem for the Laplace equation, of a form noticeably different from the Cauchy-Kowalewski theorem. We prove that if we are given the value of either $\Phi(\mathbf{x})$ or $\partial\Phi/\partial n$ at every point of a closed boundary, and the given functions satisfy the restrictive conditions imposed by the Gauss theorem, then there is a unique solution of the Laplace equation within the boundary which takes these boundary values, with the modification that if $\partial\Phi/\partial n$ is given at each point of the boundary, we may add to $\Phi(\mathbf{x})$ any arbitrary constant.

Suppose $\Phi_1(\mathbf{x})$ and $\Phi_2(\mathbf{x})$ both satisfy (6.1.1) and take the same boundary values at each point. Then

$$\Phi_3(\mathbf{x}) = \Phi_1(\mathbf{x}) - \Phi_2(\mathbf{x})$$

satisfies (6.1.1) at each point of the interior of the domain, and at each point of its boundary either $\Phi_3(\mathbf{x})$ or $\partial\Phi_3/\partial n$ vanishes. If we substitute $\Phi_3(\mathbf{x})$ both for $\Psi(\mathbf{x})$ and $\Phi(\mathbf{x})$ in (6.3.1), we deduce that

$$\int_V \{\text{grad}\,\Phi_3\}^2\,d\tau = 0,$$

where the vector

$$\text{grad}\,\Phi = \left\{\frac{\partial \Phi}{\partial x_i}\right\}.$$

The conclusion is clearly only possible if

$$\text{grad}\,\Phi_3 = 0,$$

which implies that $\Phi_3(\mathbf{x})$ is a constant. This must be zero if $\Phi_3(\mathbf{x})$ is known to be zero at any point of the boundary.

The theorem can be extended to unbounded regions of the space, provided either $\Phi(\mathbf{x})$ or $\partial\Phi/\partial n$ is known at every point of all the finite boundaries, and it is given that as $|\mathbf{x}| \to \infty$, $\Phi(\mathbf{x})$ vanishes like $|\mathbf{x}|^{-m}$, where in an odd-dimensional space $2m = n - 1$, and in an even-dimensional space $2m = n$. The proof is similar; a bounded domain is cut out of the unbounded by taking the part contained within the n-dimensional sphere $|\mathbf{x}| = R$. The theorem as proved above can be applied to the bounded domain. On the spherical surface $\Phi_3(x)\partial\Phi_3/\partial n$ is of order $(R)^{-2m-1}$, while the surface area is of order R^{n-1}, so that in either case the contribution from the spherical surface to the surface integral can be made as small as we please.

It should be noted that the conditions under which we have proved the theorem are, one boundary condition at each point of a closed boundary, whereas the Cauchy-Kowalewski theorem requires two conditions at each point of an open boundary. There is therefore a very marked divergence between the two theorems.

Although we can solve the Laplace equation, given a mixture of values of the function and its normal derivative, in a large number of physical problems the boundary data are of one type or the other. In electrostatics, for example, we may wish to find a potential function which has given fixed values on the boundaries. When therefore a solution is to be determined from values of the function given at each point of the boundary, we say that we have a Dirichlet problem. If the normal derivative is given everywhere, we say that we have a Neumann problem. Such a problem would be typical of classical hydrodynamics where the given data are the velocities of the boundaries.

6.4 The fundamental solution and its use

Let

$$r^2 = (x_i - a_i)(x_i - a_i), \tag{6.4.1}$$

where $\mathbf{a} = (a_i)$ is some fixed point.

Consider then

$$\Psi(\mathbf{x}) = r^{-m}, \tag{6.4.2}$$

so that

$$\Psi_i = -m(x_i - a_i)r^{-m-2},$$

and

$$\begin{aligned}\nabla_n{}^2\Psi &= -nmr^{-m-2} + m(m+2)(x_i - a_i)(x_i - a_i)r^{-m-4},\\ &= m(m - n + 2)r^{-m-2}.\end{aligned} \tag{6.4.3}$$

Hence if $m = n - 2$ (or when $n = 2$, $\Psi(\mathbf{x}) = \log r$), we have found a solution of (6.1.1). We shall call this the fundamental solution.

Now, as before, let V be a domain bounded by the closed surface Σ, and we derive the domain V' from V by removing the interior of the sphere $r = \delta$ from V. It is assumed that $\mathbf{a}$ is an interior point of V. Then we apply the Green's identity to the domain V', wherein we take $\Psi(\mathbf{x})$ to be the fundamental solution, and we assume that $\Phi(\mathbf{x})$ is a solution of the Poisson equation (6.1.2). Since then $\Psi(\mathbf{x})$ is analytic in V' and satisfies the Laplace equation, we see that:

$$-\Omega_n\int_{V'}\rho(\mathbf{x})\,\Psi(\mathbf{x})\,d\tau = \int_{\Sigma}\left\{\Psi(\mathbf{x})\frac{\partial\Phi}{\partial n} - \Phi(\mathbf{x})\frac{\partial\Psi}{\partial n}\right\}dS$$
$$-\int_{r=\delta}\left\{\Psi(\mathbf{x})\frac{\partial\Phi}{\partial r} - \Phi(\mathbf{x})\frac{\partial\Psi}{\partial r}\right\}dS.$$

On the surface of the sphere $r = \delta$, $\Psi(\mathbf{x})$ is of order r^{2-n} and $\partial\Psi/\partial r$ is of order r^{1-n}, while it is assumed that $\partial\Phi/\partial r$ and $\Phi(\mathbf{x})$ are analytic near $\mathbf{x} = \mathbf{a}$. Thus by the mean value theorem, if we denote an average value on the surface by a raised bar $\overline{}$

$$\int_{r=\delta}\Psi(\mathbf{x})\frac{\partial\Phi}{\partial r}\,dS = \Omega_n\delta\overline{\frac{\partial\Phi}{\partial r}}$$

while

$$\int_{r=\delta}\frac{\partial\Psi(\mathbf{x})}{\partial r}\Phi(\mathbf{x})\,dS = (2-n)\Omega_n\overline{\Phi}.$$

The singularity of the left-hand side at $\mathbf{x} = \mathbf{a}$ is integrable, and so if we let $\delta \to 0$, we deduce that

$$(n-2)\Phi(\mathbf{a}) = \int_V\rho(\mathbf{x})\,\Psi(\mathbf{x})\,d\tau - \Omega_n{}^{-1}\int_{\Sigma}\left\{\Psi(\mathbf{x})\frac{\partial\Phi}{\partial n} - \Phi(\mathbf{x})\frac{\partial\Psi}{\partial n}\right\}dS. \tag{6.4.4}$$

This can be interpreted as giving the solution of (6.1.2) at any interior point if we know the values both of the function and its normal derivative at every point on the boundary. Clearly we may derive an exactly equivalent result for the Laplace equation by putting $\rho(\mathbf{x}) = 0$. This result appears to be in conflict with the theorem we proved in Section 6.3. Since we have proved that there is a unique solution if either the function or its normal derivative is given at each point, then if both are given the solution must in general be over-determined. We would expect the boundary data to be inconsistent in the sense that, if we derived two solutions, say using the values of the function data for the first, and the values of the normal derivative for the second, we would not expect the two solutions to agree.

The resolution of the paradox is that, in using the fundamental solution in (6.4.4), we have not made the full use of the freedom that is available to us. We shall return to this question when we consider the subject of Green's functions, and we shall then show that we can get an equivalent result to (6.4.4), which is consistent with Section 6.3.

It is worth pointing out also, before we leave this section, that of the two terms on the right of (6.4.4), the first is a particular integral of (6.1.2) and the second is a solution of (6.1.1). This not only illustrates property B of Section 6.1, but also implies that we may always take the first term as a particular integral of (6.1.2) when we use this property. For this reason we shall not in general concern ourselves again with the Poisson equation.

6.5 The equation in two-dimensional Cartesian coordinates

We want now to start the examination of some solutions of the Laplace equation in two and three dimensions. In the following sections, and in subsequent chapters, we shall make extensive use of methods of expansion. The basis of what we shall be doing is the additive property, property A, of solutions of the Laplace equation, so that from sets of simple solutions we shall be able to construct more general solutions. We shall at every stage be appealing implicitly to the uniqueness theorem, so that if we can find a solution of the equation which satisfies the boundary conditions, it will be the only solution, although this does not of course prevent us from writing the same solution in a different functional form. This will justify any apparent arbitrariness in the process by which the solution is derived.

We shall also require the application of expansion theorems for some of the well-known orthogonal functions of mathematical physics. In this particular section we shall require only the ordinary Fourier theorem. A proof of this, under stated conditions which will be quite sufficient for our purpose, is given in Appendix 3. For proofs of more general expansion theorems under similar restrictions the reader is referred to [19].

Consider the Laplace equation in two dimensions in Cartesian coordinates

$$\frac{\partial^2 \Phi}{\partial x^2} + \frac{\partial^2 \Phi}{\partial y^2} = 0. \tag{6.5.1}$$

It is shown in textbooks of analysis like [20] that if $f(z)$ is a regular function of $z = x + iy$, and if

$$f(z) = u(x, y) + iv(x, y)$$

then both $u(x, y)$ and $v(x, y)$ satisfy the Laplace equation. Conversely, if $\Phi(x, y)$ has continuous second differential coefficients satisfying (6.5.1) at each interior point of a domain, then

$$u(x, y) = \frac{\partial \Phi}{\partial x}, \quad \text{and} \quad v(x, y) = \frac{\partial \Phi}{\partial y}$$

satisfy the Cauchy-Riemann conditions, and have continuous first derivatives at every interior point, and so are the real and imaginary parts of an analytic function. Hence we expect solutions of (6.5.1) to be analytic in the interior of the domain.

We look for solutions

$$\Phi(x, y) = X(x)Y(y) \tag{6.5.2}$$

where $X(x)$ is a function of x alone and $Y(y)$ is a function of y alone. We call such a solution 'separable'. Then

$$Y(y)X''(x) + X(x)Y''(y) = 0,$$

where dashes denote total differentiation with respect to the argument of each particular function. Since

$$\frac{X''(x)}{X(x)} = -\frac{Y''(y)}{Y(y)}$$

and the first is a function of x alone and the second is a function of y alone, each term separately is a constant.

If $X''(x) = 0$, then $Y''(y) = 0$ and

$$X(x) = Ax + B \quad \text{and} \quad Y(x) = Cy + D. \tag{6.5.3}$$

If $X''(x) = \nu^2 X$, then $Y'' + \nu^2 Y = 0$ and

$$X(x) = A \cosh(\nu x + \delta_\nu), \quad \text{and} \quad Y(y) = C \cos(\nu y + \varepsilon_\nu). \tag{6.5.4}$$

If we take a negative constant, the roles of $X(x)$ and $Y(y)$ are reversed.

We can therefore produce a more complicated solution by adding a set of such solutions for various values of ν:

$$\Phi(x, y) = (A_0 x + B_0)(C_0 y + D_0) + \sum_\nu \{A_\nu \cosh(\nu x + \delta_\nu) \times \cos(\nu y + \varepsilon_\nu) + B_\nu \cos(\nu x + \delta_\nu') \cosh(\nu y + \varepsilon_\nu')\}. \tag{6.5.5}$$

In this solution A_0, B_0, C_0, D_0, A_ν, B_ν, δ_ν, ε_ν, δ_ν' and ε_ν' are all constants at our disposal, and the set of values of ν over which the summation is to be made is not known.

Let us consider the solution of the following Dirichlet problem. We require a solution of (6.5.1) which vanishes on each of the lines $y = 0$,

$y = b$ and $x = a$, while on $x = 0$ it takes the values of a given function $f(y)$.

We can satisfy the boundary conditions on $y = 0$ and $y = b$ if each separate function of y vanishes at both limits. Now the problem of finding non-trivial solutions of

$$Y''(y) = \lambda Y(y)$$

where λ is a constant, which satisfy $Y(0) = Y(b) = 0$, is possibly the simplest of all eigenvalue problems associated with second-order ordinary differential equations. It can be solved only if λ takes one of the eigenvalues

$$b^2\lambda = -n^2\pi^2,$$

where n is an integer. The corresponding eigenfunction is

$$Y_n(y) = A_n \sin(n\pi y/b).$$

We can also satisfy the boundary condition on $x = a$, if

$$X_n(x) \propto \sinh\{n\pi(a - x)/b\}.$$

The complete solution is then:

$$\Phi(x, y) = \sum_{n=1}^{\infty} A_n \sinh\{n\pi(a - x)/b\} \sin(n\pi y/b). \tag{6.5.6}$$

In particular, as $x \to 0$

$$\Phi(x, y) \to \sum_{n=1}^{\infty} A_n \sinh(n\pi a/b) \sin(n\pi y/b). \tag{6.5.7}$$

Now provided $f(y)$ satisfies the right conditions, it can be expanded as a Fourier sine series in $0 < y < b$

$$f(y) = \sum_{n=1}^{\infty} c_n \sin(n\pi y/b), \tag{6.5.8}$$

where

$$bc_n = 2\int_0^b f(\eta) \sin(n\pi\eta/b)\, d\eta.$$

Since, however,

$$\Phi(0, y) = f(y),$$

and Fourier series are unique, (6.5.7) and (6.5.8) must be equal term by term, and so

$$A_n b \sinh(n\pi a/b) = 2\int_0^b f(\eta) \sin(n\pi\eta/b)\, d\eta. \tag{6.5.9}$$

It will be noticed at once that the conditions on $f(y)$ for its expansion allow a finite number of finite discontinuities, while the solution (6.5.6) is analytic in the interior of the rectangle. This last condition does not, however, imply that the solution need be analytic in the closed rectangle including the boundary. A discontinuity will be reflected in the fact that the series (6.5.6) will not converge uniformly up to the boundary.

From solutions like (6.5.6) and (6.5.9) we can build up a solution of the Dirichlet problem for the rectangle. We can consider the solution to be the sum of four functions, each itself a solution of the Laplace equation, vanishing on three of the sides, and taking the given set of values on the fourth. Each of the four parts will therefore have a form similar to that above.

If the boundary at $x = a$ is replaced by one at $x = \infty$, the corresponding form of solution will be

$$\Phi(x, y) = \sum_{n=1}^{\infty} A_n \exp\left\{-\frac{n\pi x}{b}\right\} \sin\left\{\frac{n\pi y}{b}\right\}, \tag{6.5.10}$$

where

$$b\, A_n = 2\int_0^b f(\eta) \sin (n\pi\eta/b)\, d\eta.$$

If we considered the Neumann problem, wherein $\partial\Phi/\partial y$ is to vanish on $y = 0$ and $y = b$, $\partial\Phi/\partial x$ is to vanish on $x = a$, and $\partial\Phi/\partial x = f(y)$ on $x = 0$, by an exactly similar argument we would find that

$$\Phi(x, y) = \sum_{n=1}^{\infty} B_n \cosh \{n\pi(a - x)/b\} \cos \{n\pi y/b\}, \tag{6.5.11}$$

where

$$-n\pi B_n \sinh (n\pi a/b) = 2\int_0^b f(\eta) \cos (n\pi y/b). \tag{6.5.12}$$

6.6 An example by Hadamard

Before leaving this case it is instructive to look at an example given by HADAMARD [8], which illustrates why the physical boundary conditions for the Laplace equation are not those considered in the Cauchy-Kowalewski theorem.

From (6.5.5) we see that

$$\Phi_\nu(x, y) = \nu^{-2} \cosh \nu x \cos \nu y, \tag{6.6.1}$$

is a solution of the Laplace equation. It clearly satisfies the conditions that, on $x = 0$,

$$|\, \Phi_\nu(x, y)\, | = |\, \nu^{-2} \cos \nu y\, | \leqslant \nu^{-2}$$

and

$$\frac{\partial \Phi_\nu}{\partial x} = 0.$$

On the other hand, on the line $x = x_0 > 0$ at each point

$$\nu y = k\pi,$$

for integer k,

$$|\Phi_\nu(x, y)| = \nu^{-2} \cosh \nu x_0.$$

Now given two positive numbers, ε (arbitrarily small) and M (arbitrarily large) we can choose ν so large that simultaneously

$$\nu^{-2} < \varepsilon$$

and

$$\nu^{-2} \cosh \nu x_0 > M.$$

In other words, therefore, we can find a solution of the Laplace equation, which takes values as small as we please on $x = 0$, and whose normal derivative vanishes on $x = 0$, but which takes functional values as large as we please as close as we please to $x = 0$.

Since solutions of the Laplace equation are additive, we could add to this another solution whose boundary conditions have been given on $x = 0$. Hence we conclude that if a solution of the Laplace equation is determined by giving the values of the function and its normal derivative on $x = 0$, which is what the Cauchy-Kowalewski conditions envisage, a small change in the boundary conditions can produce an indefinitely large change in the solution, as close as we please to the boundary. We say therefore that a solution determined by Cauchy-Kowalewski boundary conditions is not a continuous function of those boundary conditions

6.7 The equation in two-dimensional polar coordinates

We can alternatively consider the equation to be expressed in terms of plane polar coordinates (r, θ) where $x = r \cos \theta$, and $y = r \sin \theta$. The equation then has the form

$$\frac{1}{r}\frac{\partial}{\partial r}\left(r\frac{\partial \Phi}{\partial r}\right) + \frac{1}{r^2}\frac{\partial^2 \Phi}{\partial \theta^2} = 0. \tag{6.7.1}$$

If we proceed as before, we look for solutions

$$\Phi(r, \theta) = \rho(r)\Theta(\theta),$$

and we see that for some constant λ

$$\frac{r}{\rho}\frac{d}{dr}\left(r\frac{d\rho}{dr}\right) = -\frac{\Theta''}{\Theta} = \lambda. \tag{6.7.2}$$

As in the previous section we see that the solution for $\Theta(\theta)$ is of the form $\cosh \{\theta\sqrt{(-\lambda)} + \delta\}$, $A\theta + B$, or $\cos(\theta\sqrt{\lambda} + \delta)$, depending on whether λ is negative, zero or positive. Suppose that, as will usually be the case, we can draw round the pole $r = 0$ a simple closed curve, lying entirely within the domain in which we are seeking a solution of the equation. Then if we go once round this curve, θ will either increase or decrease by 2π, but physically we shall return to the same place in the domain. If, then, our solution of the equation is to be single-valued in the domain it must have a period of 2π. Of the three forms of solution considered above, the first cannot be periodic, the second only if $A = 0$, and the last can only have a period 2π if $\sqrt{\lambda} = n$, where n is an integer. Thus we find that

$$r\frac{d}{dr}\left(r\frac{d\rho}{dr}\right) = n^2\rho, \tag{6.7.3}$$

which gives

$$\rho(r) = Ar^n + Br^{-n}. \tag{6.7.4}$$

We shall therefore have a solution of (6.7.1)

$$\Phi(r, \theta) = A_0 + \sum_{n=1}^{\infty} (A_n r^n + B_n r^{-n}) \cos (n\theta + \delta_n). \tag{6.7.5}$$

If we expand the cosine and regroup the terms, we may write this alternatively as

$$\Phi(r, \theta) = \sum_{n=-\infty}^{\infty} r^n(\alpha_n \cos n\theta + \beta_n \sin n\theta), \tag{6.7.6}$$

where if $n \geqslant 0$

$$\alpha_n = A_n \cos \delta_n \quad \text{and} \quad \beta_n = -A_n \sin \delta_n,$$

and if $n < 0$

$$\alpha_n = B_{-n} \cos \delta_{-n} \quad \text{and} \quad \beta_n = B_{-n} \sin \delta_{-n}.$$

In this form $\Phi(r, \theta)$ is obviously the real part of the Laurent series

$$\sum_{n=-\infty}^{\infty} (\alpha_n - i\beta_n)z^n,$$

where $z = re^{i\theta}$. If the negative powers of r terminate at some value r^{-N}, the function represented by the Laurent series has a pole of order N at $r = 0$ and the function, and so its real part also, tends uniformly to infinity as $r \to 0$. If the negative powers do not terminate, the function has an essential singularity at $r = 0$, and by the Weierstrass theorem the real part takes values as large as we like as close as we please to $r = 0$. Hence, if solutions of the equation are to be bounded near $r = 0$, no negative powers of r can occur.

Now consider the Dirichlet problem for the circle. We seek first a solution of (6.7.1) inside the circle $r = a$, which takes the boundary values $\Phi(a, \theta) = f(\theta)$ on the circle. We shall assume that $f(\theta)$ is a periodic function of θ, of period 2π, which satisfies the necessary conditions so that it possesses a Fourier series in the interval $0 \leqslant \theta < 2\pi$.

Since our solution is to be finite at $r = 0$, no negative powers of r can appear in the solution, which therefore, from (6.7.6), is of the form

$$\Phi(r, \theta) = \alpha_0 + \sum_{n=1}^{\infty} r^n(\alpha_n \cos n\theta + \beta_n \sin n\theta),$$

and in particular on the circle $r = a$

$$\Phi(a, \theta) = \alpha_0 + \sum_{n=1}^{\infty} a^n(\alpha_n \cos n\theta + \beta_n \sin n\theta). \tag{6.7.7}$$

Now $f(\theta)$ has a Fourier series, by assumption,

$$f(\theta) = \frac{a_0}{2} + \sum_{n=1}^{\infty} (a_n \cos n\theta + b_n \sin n\theta),$$

where

$$\pi a_n = \int_0^{2\pi} f(\theta') \cos n\theta' \, d\theta' \quad \text{and} \quad \pi b_n = \int_0^{2\pi} f(\theta') \sin n\theta' \, d\theta',$$

and these two series must be equal term by term, by the uniqueness of Fourier series. Hence the solution we seek is

$$\pi\Phi(r, \theta) = \tfrac{1}{2}\int_0^{2\pi} f(\theta') \, d\theta' + \sum_{n=1}^{\infty} \left(\frac{r}{a}\right)^n \int_0^{2\pi} f(\theta') \cos \{n(\theta - \theta')\} \, d\theta'. \tag{6.7.8}$$

Provided $r < a$, this series converges uniformly and we may reverse the order of summation and integration, and so

$$2\pi\Phi(r, \theta) = \int_0^{2\pi} f(\theta') p\left\{\frac{r}{a}, \theta - \theta'\right\} d\theta', \tag{6.7.9}$$

where

$$\begin{aligned} p(\xi, \phi) &= 1 + 2\sum_{n=1}^{\infty} \xi^n \cos n\phi, \\ &= \mathscr{R}\left[1 + 2\sum_{n=1}^{\infty} \{\xi^n \exp(in\phi)\}\right] \\ &= \mathscr{R}\left[1 + \frac{2\xi e^{i\phi}}{1 - \xi e^{i\phi}}\right] = \mathscr{R}\left[\frac{1 + \xi e^{i\phi}}{1 - \xi e^{i\phi}}\right], \\ &= \frac{1 - \xi^2}{1 - 2\xi \cos \phi + \xi^2}. \end{aligned}$$

We thus derive the expression known as Poisson's integral

$$2\pi\Phi(r, \theta) = \int_0^{2\pi} \{(a^2 - r^2) f(\theta')\}\{a^2 - 2ar \cos(\theta - \theta') + r^2\}^{-1} \, d\theta'. \tag{6.7.10}$$

If, instead, we considered the region $r > a$, and sought a solution which was bounded at infinity and reduced to $f(\theta')$ on $r = a$, we would find a similar result with the roles of r and a reversed.

The corresponding Neumann problem would be to find a solution of (6.7.1) in $r < a$, given that on $r = a$, $\partial\Phi/\partial r = f(\theta')$. We should then compare the Fourier series for $f(\theta)$ with

$$\sum_{n=1}^{\infty} na^{n-1}(\alpha_n \cos n\theta + \beta_n \sin n\theta)$$

so that unless

$$\pi a_0 = \int_0^{2\pi} f(\theta')\, d\theta' = 0,$$

the problem cannot be solved. If we assume that this condition is satisfied, then

$$\pi\Phi(r, \theta) = a\left[\alpha_0 + \sum_{n=1}^{\infty} \frac{1}{n}\left(\frac{r}{a}\right)^n \int_0^{2\pi} f(\theta') \cos\{n(\theta - \theta')\}\, d\theta'\right].$$

The series in this case is clearly the real part of the logarithmic series

$$\sum_{n=1}^{\infty} \frac{1}{n}(\xi e^{i\phi})^n$$

and so the final answer is

$$\pi\Phi(r, \theta) = \alpha_0 a - \frac{a}{2}\int_0^{2\pi} f(\theta') \log\{(a^2 - 2ar\cos(\theta - \theta') + r^2)/a^2\}\, d\theta'. \tag{6.7.11}$$

It should be noted that the results (6.7.10) and (6.7.11) can be obtained independently as exercises in the theory of functions of a complex variable. In particular a derivation of (6.7.10) is given in [20]. The derivation is more restrictive than the one we have given above, as it assumes that the function $f(\theta)$ gives the values taken on the circle by the real part of a complex function in a region in which it is defined to be regular. Therefore, for example, $f(\theta)$ will be continuous on the circle. This is clearly imposing far greater restrictions than we need.

We may also, of course, use the theory of analytic functions in other ways to assist in finding the solution of a given problem. Possibly the most useful result in this connection is the Riemann mapping theorem, which tells us that we can map simply the interior of any bounded, simply-connected domain which is surrounded by a Jordan contour on to the interior of the unit circle. Thus formally we can reduce a very large class of problems to one of the two considered above.

6.8 The equation in three-dimensional Cartesian coordinates

In three dimensions, if Cartesian coordinates are used, the equation has the standard form:

$$\frac{\partial^2\Phi}{\partial x^2} + \frac{\partial^2\Phi}{\partial y^2} + \frac{\partial^2\Phi}{\partial z^2} = 0. \tag{6.8.1}$$

If we look for separable solutions

$$\Phi(x, y, z) = X(x)\,Y(y)\,Z(z),$$

we must have

$$X''(x) = \lambda X(x), \quad Y''(y) = \mu Y(y) \quad \text{and} \quad Z''(z) = \nu Z(z),$$

where

$$\lambda + \mu + \nu = 0.$$

In particular, a solution of the equation within the rectangular parallelepiped bounded by the planes $x = 0$, $x = a$, $y = 0$, $y = b$, $z = 0$, $z = c$, and which vanishes on every plane face excepting only $x = 0$, on which it takes the values of $f(y, z)$, is

$$\Phi(x, y, z) = \sum_{n=0}^{\infty}\sum_{m=0}^{\infty} A_{nm} \operatorname{cosech} pa \sinh\{p(a - x)\} \sin(\pi ny/b) \times \sin(\pi mz/c), \tag{6.8.2}$$

wherein

$$p^2 = \left(\frac{n^2}{b^2} + \frac{m^2}{c^2}\right)\pi^2$$

and

$$bcA_{nm} = 4\int_0^b d\eta \int_0^c f(\eta, \zeta) \sin(\pi n\eta/b) \sin(\pi m\zeta/c)\, d\zeta. \tag{6.8.3}$$

We may then build up solutions of the Dirichlet problem for the rectangular parallelepiped as the sum of six solutions having a form similar to (6.8.2). In principle, therefore, this is no different from the equivalent problem in two dimensions.

6.9 The equation in spherical polar coordinates

We now examine the equation when it is expressed in terms of the spherical polar coordinates R, θ and ψ, wherein

$$x = R\sin\theta\cos\psi, \quad y = R\sin\theta\sin\psi \quad \text{and} \quad z = R\cos\theta.$$

The equation takes the form

$$R^{-2}\left\{\frac{\partial}{\partial R}\left(R^2\frac{\partial\Phi}{\partial R}\right) + \operatorname{cosec}\theta\frac{\partial}{\partial\theta}\left(\sin\theta\frac{\partial\Phi}{\partial\theta}\right) + \operatorname{cosec}^2\theta\frac{\partial^2\Phi}{\partial\psi^2}\right\} = 0. \tag{6.9.1}$$

When we substitute

$$\Phi(R, \theta, \psi) = \rho(R)\, \Theta(\theta)\, \Psi(\psi)$$

we see first that we shall require that

$$\Psi''(\psi) = \lambda\Psi(\psi).$$

By repeating the argument that we used in Section 6.7, we see that if this is to be a single-valued function of ψ we must have

$$\lambda = -m^2,$$

where m is an integer. We next deduce that for some constant c

$$\operatorname{cosec}\theta\frac{d}{d\theta}\left\{\sin\theta\frac{d\Theta}{d\theta}\right\} + (c - m^2\operatorname{cosec}^2\theta)\Theta = 0. \tag{6.9.2}$$

If $c \geqslant -\frac{1}{4}$ there is no loss of generality in making the substitution

$$c = \nu(\nu + 1).$$

If, further, we make the usual change of variable $\mu = \cos\theta$, the equation takes the form

$$\frac{d}{d\mu}\left\{(1 - \mu^2)\frac{d\Theta}{d\mu}\right\} + \left\{\nu(\nu + 1) - \frac{m^2}{1 - \mu^2}\right\}\Theta = 0, \tag{6.9.3}$$

which is the standard form of the equation for the associated Legendre functions [24]. When $m = 0$ the equation reduces to that for the ordinary Legendre functions. The solution is written

$$\Theta(\mu) = AP_\nu{}^m(\mu) + BQ_\nu{}^m(\mu). \tag{6.9.4}$$

We prove in Appendix 4 that almost all solutions of this equation when $m = 0$ have logarithmic singularities when $\mu = \pm 1$. The only solution regular at $\mu = \pm 1$ is $P_\nu{}^0(\mu) = P_\nu(\mu)$, and then only if $\nu = n$, an integer. An exactly similar argument applies to $P_\nu{}^m(\mu)$. Hence if our solution is to be bounded at any points on the half-lines $\theta = 0$ and $\theta = \pi$, we are restricted to solutions like $P_n{}^m(\mu)$. When $m = 0$, which means that our solution is independent of ψ and so has axial symmetry, the solution is $P_n(\mu)$, the ordinary Legendre polynomial.

In either case this gives an equation for $\rho(R)$:

$$\frac{d}{dR}\left\{R^2\frac{d\rho}{dR}\right\} = n(n + 1)\rho(R), \tag{6.9.5}$$

which has the solution

$$\rho(R) = CR^n + DR^{-n-1}. \tag{6.9.6}$$

If $c < \frac{1}{4}$, ν is complex, and it may be similarly shown that no regular solution exists. If then we wish to construct a solution which is both single-valued and analytic in a three-dimensional region including the pole $R = 0$, it will have the form

$$\Phi(R, \theta, \psi) = \sum_{n=0}^{\infty} \sum_{m=0}^{n} (C_{nm}R^n + D_{nm}R^{-n-1})P_n^m(\cos\theta)\cos(m\psi + \delta_{nm}), \tag{6.9.7}$$

or, if the solution is to have axial symmetry

$$\Phi(R, \theta) = \sum_{n=0}^{\infty} \{C_n R^n + D_n R^{-n-1}\} P_n(\cos\theta). \tag{6.9.8}$$

We shall consider the Dirichlet problem for the sphere. Let us suppose first that we have axial symmetry, so that the problem is to find a solution of (6.9.1) inside the sphere $R = a$, which for all ψ in the range $0 \leqslant \psi < 2\pi$, takes the value $f(\theta)$ on $R = a$. Then, since our solution must remain finite as $R \to 0$, it is of the form

$$\Phi(R, \theta) = \sum_{n=0}^{\infty} C_n R^n P_n(\cos\theta). \tag{6.9.9}$$

We now assume that $f(\theta)$ can be expressed as an infinite series of Legendre polynomials

$$f(\theta) = \sum_{n=0}^{\infty} b_n P_n(\cos\theta). \tag{6.9.10}$$

If we make use of the orthogonality relation for the Legendre polynomials, we deduce that

$$2b_n = (2n + 1)\int_0^{\pi} f(\theta')P_n(\cos\theta')\sin\theta'\, d\theta'. \tag{6.9.11}$$

When this substitution is made into (6.9.10), and the expansion is compared to (6.9.9), we see that

$$2\Phi(R, \theta) = \int_0^{\pi} f(\theta')\left\{\sum_{n=0}^{\infty}\left(\frac{R}{a}\right)^n P_n(\cos\theta)P_n(\cos\theta')\right\} d\theta'. \tag{6.9.12}$$

If we do not have axial symmetry, we must assume that on $R = a$ the solution takes the values $f(\theta, \psi)$, and this function must be periodic of period 2π in ψ. The expansion must have the form

$$\Phi(R, \theta, \psi) = \sum_{n=0}^{\infty}\sum_{m=0}^{n}\left(\frac{R}{a}\right)^n P_n^m(\cos\theta)(\alpha_{nm}\cos m\psi + \beta_{nm}\sin m\psi). \tag{6.9.13}$$

To find the coefficients in this expansion we must use not only the orthogonality relation for the associated Legendre functions [24], but also those for the circular functions. We may then show that

$$2\pi\{(n+m)!\}\alpha_{nm} = (2n+1)\{(n-m)!\} \times \int_0^{\pi} d\theta' \sin\theta' P_n^m(\cos\theta') \int_0^{2\pi} f(\theta', \psi') \cos m\psi' \, d\psi', \tag{6.9.14}$$

and

$$2\pi\{(n+m)!\}\beta_{nm} = (2n+1)\{(n-m)!\} \times \int_0^{\pi} d\theta' \sin\theta' P_n^m(\cos\theta') \int_0^{2\pi} f(\theta', \psi') \sin m\psi' \, d\psi'. \tag{6.9.15}$$

If then we substitute these coefficients into (6.9.13) we may express the solution in a form similar to (6.7.9), namely

$$2\pi\Phi(R, \theta, \psi) = \int_0^{\pi} d\theta' \sin\theta' \int_0^{2\pi} f(\theta', \psi') p(R/a, \theta, \theta', \psi - \psi') \, d\psi', \tag{6.9.16}$$

where

$$p\{\xi, \theta, \theta', \phi\} = \sum_{n=0}^{\infty} (2n+1)\xi^n \left\{ \sum_{m=0}^{n} \frac{\{(n-m)!\}}{\{(n+m)!\}} P_n^m(\cos\theta) P_n^m(\cos\theta') \cos m\phi \right\}.$$

Now if we use the addition theorem for the Legendre polynomials [24], the sum contained in the curly brackets is just

$$\tfrac{1}{2} P_n\{\cos\Theta\},$$

where

$$\cos\Theta = \cos\theta\cos\theta' + \sin\theta\sin\theta'\cos\phi, \tag{6.9.17}$$

and so is the cosine of the angle contained between unit vectors whose polar Eulerian angles are θ and θ', and whose azimuth angles differ by ϕ.

We have, in addition, the identity that

$$(1-\xi^2)(1 - 2\xi\cos\Theta + \xi^2)^{-3/2} = \sum_{n=0}^{\infty} (2n+1)\xi^n P_n(\cos\Theta).$$

Finally therefore we get the Poisson integral for the sphere

$$4\pi\Phi(R, \theta, \psi) = (a^2 - R^2)a \int_0^{\pi} d\theta' \sin\theta' \int_0^{2\pi} f(\theta', \psi') \times \{a^2 - 2aR\cos\Theta + R^2\}^{-3/2} \, d\psi', \tag{6.9.18}$$

where the angle Θ is defined by (6.9.17).

As in the case of the circle, we can easily deduce the equivalent result for the region $R > a$, and we can also find the solution for the corresponding Neumann problem by the same method.

We could approach the separation of variables in this case in another way. If we note that the differential operator in (6.9.1) is uniformly of dimension -2 in R, we see at once that there must exist solutions like

$$R^{\nu}S_{\nu}(\theta, \psi). \tag{6.9.19}$$

Then $S_{\nu}(\theta, \psi)$ is called a spherical harmonic of order ν, and satisfies the partial differential equation

$$\operatorname{cosec}\theta\frac{\partial}{\partial\theta}\left\{\sin\theta\frac{\partial S_{\nu}}{\partial\theta}\right\} + \operatorname{cosec}^2\theta\frac{\partial^2 S_{\nu}}{\partial\psi^2} + \nu(\nu+1)S = 0. \tag{6.9.20}$$

It is clear that (6.9.20) is unchanged if we write $-(1+\nu)$ for ν, and so the spherical harmonics of order ν and $-(1+\nu)$ are identical.

6.10 The equation in cylindrical polar coordinates

We finally wish to consider solutions of the Laplace equation in cylindrical polar coordinates r, θ, z, with, again, $x = r\cos\theta, y = r\sin\theta$. This does not of course exhaust the systems of orthogonal curvilinear coordinates which commonly occur in mathematical physics. In three dimensions we have not even the standard techniques of conformal transformation to reduce other problems to one which we have considered in this chapter. What we have tried to do is to illustrate how standard solutions are built up. For a detailed investigation of some other coordinate systems the reader is referred to [23] or [24].

In cylindrical polar coordinates the equation has the form

$$\frac{1}{r}\frac{\partial}{\partial r}\left\{r\frac{\partial\Phi}{\partial r}\right\} + \frac{1}{r^2}\frac{\partial^2\Phi}{\partial\theta^2} + \frac{\partial^2\Phi}{\partial z^2} = 0. \tag{6.10.1}$$

We make our usual substitution

$$\Phi(r, \theta, z) = \rho(r)\,\Theta(\theta)\,Z(z). \tag{6.10.2}$$

We observe at once that this will require that

$$\Theta'' = \lambda\Theta(\theta), \tag{6.10.3}$$

and if as usual we insist that our solution be single-valued, then $\Theta(\theta)$ must have a period in θ of 2π, and so

$$\lambda = -n^2, \tag{6.10.4}$$

where n is an integer. If $n = 0$ we have a solution independent of θ, and so with axial symmetry.

We shall also require that

$$Z''(z) = \nu Z(z). \tag{6.10.5}$$

There is no interest in the case $\nu = 0$, since that will merely reduce our problem to the one already considered in Section 6.7. Hence there are two possibilities.

If $\nu = \lambda^2$,

$$Z(z) = Ae^{\lambda z} + Be^{-\lambda z}, \tag{6.10.6}$$

while if $\nu = -\lambda^2$,

$$Z(z) = A\cos\lambda z + B\sin\lambda z. \tag{6.10.7}$$

Corresponding to (6.10.6) we find an equation for $\rho(r)$,

$$r^2\frac{d^2\rho}{dr^2} + r\frac{d\rho}{dr} + (\lambda^2 r - n^2)\rho(r) = 0, \tag{6.10.8}$$

and corresponding to (6.10.7) is the equation

$$r^2\frac{d^2\rho}{dr^2} + r\frac{d\rho}{dr} - (\lambda^2 r^2 + n^2)\rho(r) = 0. \tag{6.10.9}$$

Both of these are Bessel differential equations; the first is the equation for the ordinary Bessel functions of order n, while the second equation is for the modified functions of order n. For a detailed treatment of the Bessel functions, the reader is referred to [22]. For our immediate purposes we want properties which we shall describe.

The function

$$f(\xi) = J_\nu(\xi) = \left(\frac{\xi}{2}\right)^\nu \sum_{r=0}^{\infty} \frac{(-1)^r(\xi/2)^{2r}}{\Gamma(\nu + r + 1)\,\Gamma(r + 1)}, \tag{6.10.10}$$

is a solution of the Bessel equation of order ν,

$$\xi^2\frac{d^2f}{d\xi^2} + \xi\frac{df}{d\xi} + (\xi^2 - \nu^2)f = 0.$$

If ν is not an integer, an independent second solution of the equation is given by $J_{-\nu}(\xi)$, where we put $-\nu$ for ν in (6.10.10). When $\nu = n$, an integer, this just reduces to $(-1)^n$ times the first solution, and so we define the second independent solution:

$$Y_n(\xi) = \lim_{\nu\to n} \{J_\nu(\xi)\cos\nu\pi - J_{-\nu}(\xi)\}\operatorname{cosec}\nu\pi. \tag{6.10.11}$$

Near $\xi = 0$

$$J_n(\xi) \text{ is } O(\xi^n), \text{ and if } n \geqslant 1,\ Y_n(\xi) \text{ is } O(\xi^{-n}),$$

while $Y_0(\xi)$ is $O(\log\xi)$.

For very large real ξ

$$J_n(\xi) \sim \left(\frac{2}{\pi\xi}\right)^{1/2} \cos\left\{\xi - \frac{n\pi}{2} - \frac{\pi}{4}\right\},$$

and

$$Y_n(\xi) \sim \left(\frac{2}{\pi\xi}\right)^{1/2} \sin\left\{\xi - \frac{n\pi}{2} - \frac{\pi}{4}\right\}.$$

In a corresponding way the function

$$f(\xi) = I_\nu(\xi) = J_\nu(i\xi)\exp(-i\nu\pi/2) = \left(\frac{\xi}{2}\right)^\nu \sum_{r=0}^{\infty} \frac{(\xi/2)^{2r}}{\Gamma(\nu + r + 1)\,\Gamma(r + 1)} \tag{6.10.12}$$

is a solution of the modified equation

$$\xi^2\frac{d^2f}{d\xi^2} + \xi\frac{df}{d\xi} - (\xi^2 + \nu^2)f = 0,$$

which is $O(\xi^\nu)$ when ξ is very small, and when ξ is large and real

$$I_\nu(\xi) \sim \left(\frac{1}{2\pi\xi}\right)^{1/2} e^{\xi}.$$

When $\nu = n$ an integer, we can form an independent second solution as above:

$$K_n(\xi) = \frac{\pi}{2}\lim_{\nu\to n}\{I_{-\nu}(\xi) - I_\nu(\xi)\}\operatorname{cosec}\nu\pi, \tag{6.10.13}$$

and for small ξ, $K_n(\xi)$ behaves like $Y_n(\xi)$, and for very large ξ

$$K_n(\xi) \sim \left(\frac{2}{\pi\xi}\right)^{1/2} e^{-\xi}. \tag{6.10.14}$$

First let us consider a Dirichlet problem for the interior of the cylinder $0 < r < a$, $0 < z < z_0$. To make matters even simpler, we will assume initially that the boundary conditions have axial symmetry, and therefore the solution is independent of θ. We may then put

$$\Phi(r, z) = \Phi_1(r, z) + \Phi_2(r, z) + \Phi_3(r, z) \tag{6.10.15}$$

where $\Phi_1(r, z)$ is the solution inside the cylinder, which vanishes on $z = z_0$ and on $r = a$, and reduces to $f(r)$ on $z = 0$; $\Phi_2(r, z)$ vanishes on $z = 0$ and $r = a$, and reduces to $g(r)$ on $z = z_0$, and $\Phi_3(r, z)$ vanishes on $z = 0$ and $z = z_0$ and equals $h(z)$ on the cylindrical surface $r = a$. Since our solution is to be finite when $r = 0$, there can be no term $Y_0(\lambda r)$ or $K_0(\lambda r)$.

We can therefore satisfy the boundary conditions for $\Phi_1(r, z)$ by an expansion

$$\Phi_1(r, z) = \sum_{m=1}^{\infty} A_m J_0\left(\frac{\alpha_m r}{a}\right)\sinh\left\{\frac{\alpha_m}{a}(z_0 - z)\right\}, \tag{6.10.16}$$

where the infinite set (α_m) consists of the roots of the transcendental equation

$$J_0(\alpha) = 0.$$

It follows from the well-known comparison theorem applied to ordinary differential equations (see [11]) that as m increases these roots are separated asymptotically by π.

We may find the coefficients A_m by comparing the values taken by (6.10.16) on the plane $z = 0$, with the Fourier Bessel expansion for $f(r)$. By methods similar to those used to deduce the results for the Hankel transform in Appendix 3, we may show that

$$f(r) = \sum_{m=1}^{\infty} a_m \mathcal{J}_0\left(\frac{\alpha_m r}{a}\right), \tag{6.10.17}$$

where

$$\frac{a}{2}\{\mathcal{J}_0'(\alpha_m)\}^2 a_m = \int_0^a r f(r) \mathcal{J}_0\left(\frac{\alpha_m r}{a}\right) dr. \tag{6.10.18}$$

Hence

$$a_m = A_m \sinh\left(\frac{\alpha_m z_0}{a}\right),$$

which may be substituted into (6.10.16).

In an exactly similar way

$$\Phi_2(r, z) = \sum_{m=1}^{\infty} B_m \mathcal{J}_0\left(\frac{\alpha_m r}{a}\right) \sinh\left(\frac{\alpha_m z}{a}\right), \tag{6.10.19}$$

where

$$\frac{a}{2}\{\mathcal{J}_0'(\alpha_m)\}^2 \sinh\left(\frac{\alpha_m z_0}{a}\right) B_m = \int_0^a r g(r) \mathcal{J}_0\left(\frac{\alpha_m r}{a}\right) dr. \tag{6.10.20}$$

Finally we can satisfy the conditions for $\Phi_3(r, z)$ with an expression

$$\Phi_3(r, z) = \sum_{m=1}^{\infty} C_m I_0\left(\frac{m\pi r}{z_0}\right) \sin\left(\frac{m\pi z}{z_0}\right), \tag{6.10.21}$$

where the coefficients C_m can be found by comparison with the Fourier sine series for $h(z)$

$$z_0 C_m I_0\left(\frac{m\pi a}{z_0}\right) = 2\int_0^{z_0} h(z) \sin\left(\frac{m\pi z}{z_0}\right) dz. \tag{6.10.22}$$

If the cylinder were semi-infinite ($z_0 = \infty$), the boundary condition on z_0 would usually be replaced by $\Phi(r, z) \to 0$ as $z \to \infty$. This would require, of course, that in the boundary condition $h(z) \to 0$ as $z \to \infty$.

The solution for $\Phi_1(r, z)$ would then be replaced by

$$\Phi_1(r, z) = \sum_{m=1}^{\infty} A_m \mathcal{J}_0\left(\frac{\alpha_m r}{a}\right) \exp\left\{-\frac{\alpha_m z}{a}\right\}, \tag{6.10.23}$$

where in a similar way to that above

$$\frac{a}{2}\{\mathcal{J}_0(\alpha_m)\}^2 A_m = \int_0^a r f(r) \mathcal{J}_0\left(\frac{\alpha_m r}{a}\right) dr. \tag{6.10.24}$$

The part of the solution given by $\Phi_2(r, z)$ vanishes identically in this case. When we consider the form of expansion appropriate to $\Phi_3(r, z)$ we see that, now that $z_0 = \infty$, there is no eigenvalue problem to determine a discrete spectrum of eigenvalues similar to the set $m\pi/z_0$, over which the summation was taken in the previous example. We must therefore assume that all values are possible, and so the summation is replaced by an integration

$$\Phi_3(r, z) = \int_0^\infty C(\lambda) I_0(\lambda r) \sin \lambda z \, d\lambda. \tag{6.10.25}$$

We must now compare this with the inverse Fourier sine transform for $h(z)$.

$$h(z) = \int_0^\infty c(\lambda) \sin \lambda z \, d\lambda, \tag{6.10.26}$$

where

$$c(\lambda) = \frac{2}{\pi}\int_0^\infty h(\zeta) \sin \lambda\zeta \, d\zeta. \tag{6.10.27}$$

If we compare (6.10.25) and (6.10.26) we see that

$$\pi I_0(\lambda a)\, C(\lambda) = 2\int_0^\infty h(\zeta) \sin \lambda\zeta \, d\zeta. \tag{6.10.28}$$

If we consider solutions in the infinite slab $0 \leqslant r < \infty$, $0 \leqslant z \leqslant z_0$, then the contribution from $\Phi_3(r, z)$ vanishes, and the summations in the solutions for $\Phi_1(r, z)$ and $\Phi_2(r, z)$ are replaced by integrations. The unknown functions, corresponding to the coefficients in the series which appears in the integrand, can then be found by appealing to the inversion theorem for the Hankel transform (see Appendix 3).

We could also find solutions appropriate to the outside of the cylinder $a < r < \infty$, $0 \leqslant z \leqslant z_0$, but the eigenvalue problems involved in choosing the right form of expansion are now much more difficult, since we are no longer justified in excluding the solutions $Y_0(\lambda r)$ or $K_0(\lambda r)$.

There is no major extra complication involved if we relax the restriction on axial symmetry. If, for example, we consider again the problem of finding a solution corresponding to $\Phi_1(r, \theta, z)$, which takes the value $f(r, \theta)$ on $z = 0$, and vanishes on $r = a$, and $z = z_0$, where it is assumed that $f(r, \theta)$ is periodic in θ of period 2π, then

$$\begin{aligned}\Phi_1(r, \theta, z) = {} & \tfrac{1}{2}\sum_{m=1}^{\infty} A_{0m} J_0\left(\frac{\alpha_m r}{a}\right) \sinh\left\{\frac{\alpha_{0m}}{a}(z_0 - z)\right\} \\ & + \sum_{n=1}^{\infty}\sum_{m=1}^{\infty} (A_{nm}\cos n\theta + B_{nm}\sin n\theta) \\ & \times J_n\left(\frac{\alpha_{nm} r}{a}\right) \sinh\left\{\frac{\alpha_{nm}}{a}(z_0 - z)\right\},\end{aligned} \tag{6.10.29}$$

where

$$\pi a A_{nm}\{J_n'(\alpha_{nm})\}^2 = 2\int_0^{2\pi} d\theta \cos n\theta \int_0^a rf(r, \theta) J_n\left(\frac{\alpha_{nm} r}{a}\right) dr \tag{6.10.30}$$

and

$$\pi a B_{nm}\{J_n'(\alpha_{nm})\}^2 = 2\int_0^{2\pi} d\theta \sin n\theta \int_0^a rf(r, \theta) J_n\left(\frac{\alpha_{nm} r}{a}\right) dr, \tag{6.10.31}$$

and for every fixed n, α_{nm} form the set of positive real zeros of $J_n(\xi)$.

Exercises

6.1. Show that a function $u(x, y)$ which satisfies the Laplace equation inside the rectangle $0 \leqslant x \leqslant \pi$, $0 < y < a$, and takes the boundary values

$$u(0, y) = u(\pi, y) = 0, \qquad 0 \leqslant y \leqslant a,$$

$$u(x, a) = 0, \qquad 0 \leqslant x \leqslant \pi$$

and

$$u(x, 0) = x(\pi - x)$$

may be written

$$\pi u(x, y) = 8 \sum_{n=0}^{\infty} (2n + 1)^{-3} \operatorname{cosech} \{(2n + 1)a\} \sin \{(2n + 1)x\}$$
$$\times \sinh \{(2n + 1)(a - y)\}.$$

6.2. Find the value of a function u, at a point with polar coordinates $r = r_0 < 1$, $\theta = 0$, if u satisfies the Laplace equation inside the unit circle $r = 1$, and on $r = 1$, $u = 1$, if $0 \leqslant |\theta| \leqslant \pi/2$, and $u = 0$ if $\pi/2 < |\theta| \leqslant \pi$.

6.3. Find a solution of the Laplace equation within the cylinder $0 \leqslant z \leqslant z_0$, $0 \leqslant r \leqslant a$, if the solution vanishes on the plane $z = z_0$ and the cylindrical surface $r = a$, and takes the value $a^2 - r^2$ on the plane $z = 0$.

6.4. Find a solution of the Laplace equation in the infinite block bounded by the planes $z = 0$ and $z = z_0$, if the solution vanishes on $z = z_0$, and is equal to $\exp \{-(x^2 + y^2)^{1/2}\}$ on $z = 0$.

6.5. Express the Laplace equation in terms of the spheroidal coordinates $x = \sinh \xi \sin \eta \cos \zeta$, $y = \sinh \xi \sin \eta \sin \zeta$, $z = \cosh \xi \cos \eta$ and find the differential equations to be satisfied in a separable solution.

Answers

6.2. $$\frac{2}{\pi}\tan^{-1}\left\{\frac{1 + r_0}{1 - r_0}\right\}.$$

6.3. $$8a^3 \sum_{n=1}^{\infty} \alpha_n^{-4}\{J_0'(\alpha_n)\}^{-1} \operatorname{cosech}\left\{\frac{\alpha_n z_0}{a}\right\} J_0\left\{\frac{\alpha_n r}{a}\right\} \sinh\left\{\frac{\alpha_n}{a}(z_0 - z)\right\}$$

where α_n are the roots of $J_0(\alpha) = 0$.

6.4. $$\int_0^{\infty} (1 + \alpha^2)^{-3/2} \operatorname{cosech} (\alpha z_0) J_0(\alpha r) \sinh \{\alpha(z_0 - z)\}\, d\alpha.$$

6.5.

$$\frac{2}{\sinh \xi \sin \eta\,(\cosh 2\xi - \cos 2\eta)}\left\{\sin \eta \frac{\partial}{\partial \xi}\left(\sinh \xi \frac{\partial \Phi}{\partial \xi}\right) + \sinh \xi \frac{\partial}{\partial \eta}\left(\sin \eta \frac{\partial \Phi}{\partial \eta}\right)\right\}$$

$$+ \frac{1}{\sinh^2 \xi \sin^2 \eta} \frac{\partial^2 \Phi}{\partial \zeta^2} = 0.$$

The equations are, if $\Phi = X(\xi)\, Y(\eta)\, Z(\zeta)$

(*a*) $\dfrac{d^2 Z}{d\zeta^2} = \lambda Z$

(*b*) $\dfrac{d}{d\xi}\left(\sinh \xi \dfrac{dX}{d\xi}\right) + \left(\mu \sinh \xi - \dfrac{\lambda}{\sinh \xi}\right) X = 0$

(*c*) $\dfrac{d}{d\eta}\left(\sin \eta \dfrac{dY}{d\eta}\right) - \left(\mu \sin \eta + \dfrac{\lambda}{\sin \eta}\right) Y = 0.$

CHAPTER 7

The Wave Equation

7.1 The physical significance

Just as the Laplace equation was found to be the most fundamental of elliptic linear partial differential equations, so the most fundamental equation of normal hyperbolic type is the wave equation,

$$\frac{\partial^2 \Phi}{\partial {x_n}^2} = {\nabla_{n-1}}^2 \Phi. \tag{7.1.1}$$

In a large number of physical applications, the space coordinate x_n can be written:

$$x_n = ct, \tag{7.1.2}$$

where t is the time measured from some given instant, and c is a constant velocity representing some physical property of the system. We shall, throughout this chapter, make the substitution (7.1.2). Thus, what we shall here call the n-dimensional wave equation is, in the sense of Chapter 4, the wave equation in $n + 1$ variables. We shall make a similar usage in the next chapter.

The equation arises in the following way. Suppose we have some scalar physical quantity ϕ, defined at every point of a connected region, and the rate of transport of this quantity through the region is given by the vector $\mathbf{U}$, then a continuity equation will result:

$$\frac{\partial \phi}{\partial t} + \operatorname{div} \mathbf{U} = 0. \tag{7.1.3}$$

An obvious example is the density of a continuous fluid. If $\phi = \rho$ the fluid density, $\mathbf{U} = \rho \mathbf{v}$ where $\mathbf{v}$ is the fluid velocity.

To (7.1.3) we must append some other equation, characteristic of the physical system. If this takes the form

$$\operatorname{grad} \phi = -\frac{1}{c^2}\frac{\partial \mathbf{U}}{\partial t}, \tag{7.1.4}$$

then by elimination we get

$$\frac{\partial^2 \phi}{\partial t^2} = c^2 \nabla^2 \phi. \tag{7.1.5}$$

7.2 The wave equation in one dimension

Let us begin with the one-dimensional equation

$$\frac{\partial^2 \Phi}{\partial t^2} = c^2 \frac{\partial^2 \Phi}{\partial x^2}. \tag{7.2.1}$$

We can obtain solutions of this equation in two ways. Firstly we can continue the methods we were using in the previous chapter and look for a separable solution.

Put

$$\Phi(t, x) = \tau(t) X(x),$$

so that

$$\frac{\tau''(t)}{\tau(t)} = c^2 \frac{X''(x)}{X(x)}. \tag{7.2.2}$$

This gives solutions in the form

$$\Phi(t, x) = \sum_\lambda A_\lambda \cos(\lambda x + \varepsilon_\lambda) \cos(\lambda ct + \delta_\lambda), \tag{7.2.3}$$

where the summation is over some set of values of λ, which will be determined by the eigenvalue problem, and A_λ, ε_λ and δ_λ are constants to be fixed by the initial and boundary conditions. This form of the solution is ascribed to Bernoulli.

To obtain the second form of the solution, we observe that the characteristics of (7.2.1) are parallels to the line pair

$$x^2 - c^2 t^2 = 0. \tag{7.2.4}$$

This suggests that we use the characteristic coordinates

$$\xi = \tfrac{1}{2}(x - ct) \text{ and } \eta = \tfrac{1}{2}(x + ct). \tag{7.2.5}$$

Then the equation reduces to

$$\frac{\partial^2 \Phi}{\partial \xi \, \partial \eta} = 0, \tag{7.2.6}$$

and in this form clearly has the solution

$$\Phi(t, x) = f(2\xi) + g(2\eta) = f(x - ct) + g(x + ct), \tag{7.2.7}$$

where $f(x)$ and $g(x)$ are arbitrary functions to be found from the given initial and boundary data. In this form the solution is ascribed to D'Alembert.

It is of interest to notice that (7.2.7) represents the superposition of two 'wave trains', $f(x - ct)$ and $g(x + ct)$. The section of the wave $f(x - ct)$ which was taken at $x = x_0$ at $t = 0$ has moved to $x_0 + ct_0$ at $t = t_0$, and so $f(x - ct)$ represents a wave train steadily advancing with speed c. Similarly $g(x + ct)$ represents a retrogressive wave, retiring with speed c. If we take the progressive wave solution alone, then it is clear that the effect felt at time t at a point x was the same effect felt at $x - ct_0$ at a time t_0 earlier. This is a simple example of the Huyghens principle.

7.3 The finite string

We shall first consider in detail solutions on a bounded range of x ('finite strings'). The finite base can be expected to give rise to an obvious eigenvalue problem, and so we may anticipate that the Bernoulli solution will be convenient.

We seek a solution in the domain $t > 0$, $-l < x < l$ given the boundary conditions $\Phi(t, -l) = \Phi(t, l) = 0$, $t > 0$, and the initial conditions $\Phi(0, x) = \phi_0(x)$, $[\partial/\partial t\{\Phi(t, x)\}]_{t=0} = \phi_1(x)$ (finite string fixed at $x = \pm l$). We shall assume that $\phi_0(-l) = \phi_1(-l) = \phi_0(l) = \phi_1(l) = 0$.

A solution satisfying the boundary conditions is:

$$\Phi(t, x) = \sum_{n=0}^{\infty} \left\{ a_n \cos\left(\frac{2n+1}{2}\frac{ct\pi}{l}\right) + b_n \sin\left(\frac{2n+1}{2}\frac{ct\pi}{l}\right)\right\} \cos\left\{\frac{2n+1}{2}\frac{\pi x}{l}\right\}. \tag{7.3.1}$$

Thus from the initial conditons

$$\phi_0(x) = \sum_{n=0}^{\infty} a_n \cos\left(\frac{2n+1}{2}\frac{\pi x}{l}\right) \tag{7.3.2}$$

and

$$\phi_1(x) = \frac{c\pi}{2l}\sum_{n=0}^{\infty}(2n+1)b_n \cos\left(\frac{2n+1}{2}\frac{\pi x}{l}\right). \tag{7.3.3}$$

The coefficients a_n and b_n may then be found by simple Fourier analysis.

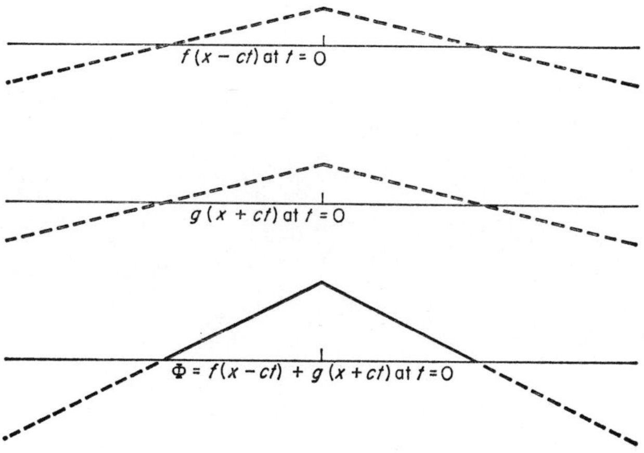

Fig. 7.1

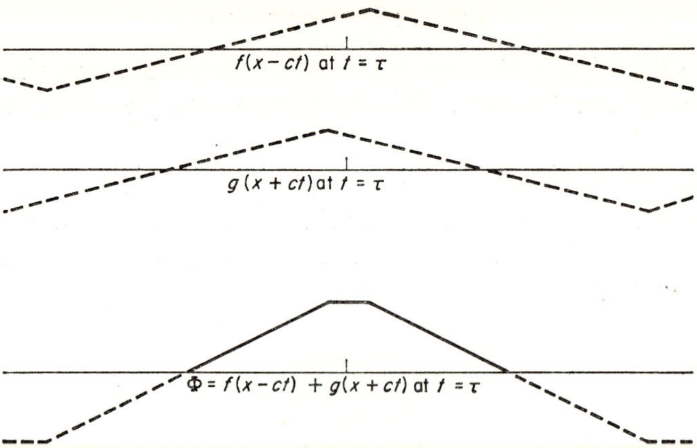

Fig. 7.2

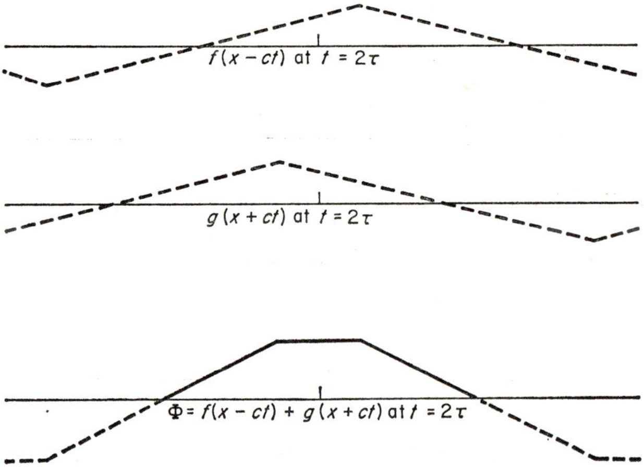

Fig. 7.3

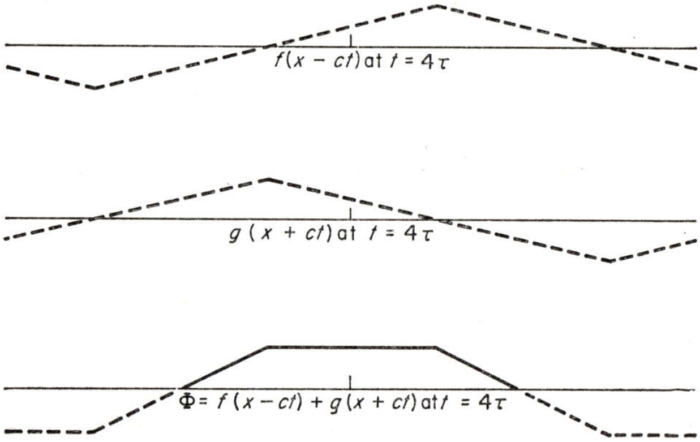

Fig. 7.4

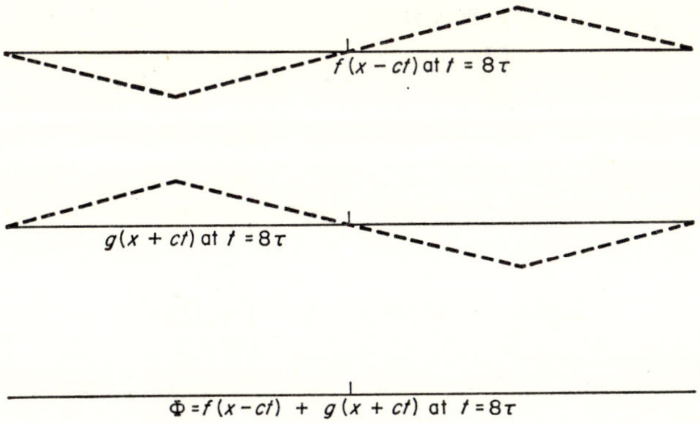

Fig. 7.5

Other boundary conditions can be dealt with in a similar way.

The obvious disadvantage of having the solution in the form (7.3.1) is that it is very difficult actually to interpret the solution. In simple physical terms, if (7.3.1) does represent a wave on a plucked string, then it is not easy to find from it the shape of the actual wave form at instants after $t = 0$. In some simple cases we may use the D'Alembert form (7.2.7), combined with a little geometric construction to do this. The method is illustrated in Figs. 7.1–7.5, where we have taken $\phi_1(x) \equiv 0$, and $\phi_0(x) = l + x$, $x < 0$ and $\phi_0(x) = l - x$, if $x > 0$.

7.4 The infinite string

In the case when x is unbounded, both above and below, there will in general be no obvious eigenvalue problem, and it will be more convenient to proceed directly from (7.2.7). We now seek a solution in $t > 0$, when $-\infty < x < \infty$. Again we assume that initially

$$\Phi(0, x) = \phi_0(x) \quad \text{and} \quad [\partial/\partial t\{\Phi(t, x)\}]_{t=0} = \phi_1(x),$$

but we now also assume that $\phi_1(x)$ is integrable over every finite interval.

Since

$$\Phi(x, t) = f(x - ct) + g(x + ct),$$

then

$$\frac{\partial}{\partial t}\Phi(x, t) = -cf'(x - ct) + cg'(x + ct), \tag{7.4.1}$$

where the prime denotes total differentiation with respect to its argument.

Hence

$$\phi_0(x) = f(x) + g(x), \tag{7.4.2}$$

and from (7.4.1)

$$\phi_1(x) = -cf'(x) + cg'(x). \tag{7.4.3}$$

If we integrate (7.4.3) we derive

$$A + \int_a^x \phi_1(\xi)\, d\xi = -cf(x) + cg(x), \tag{7.4.4}$$

where A and a are arbitrary constants. Hence

$$2cf(x) = c\phi_0(x) - A - \int_a^x \phi_1(\xi)\, d\xi, \tag{7.4.5}$$

and

$$2cg(x) = c\phi_0(x) + A + \int_a^x \phi_1(\xi)\, d\xi. \tag{7.4.6}$$

When we substitute these results back into (7.2.7) we see that

$$2c\Phi(x, t) = c\{\phi_0(x + ct) + \phi_0(x - ct)\} + \int_{x-ct}^{x+ct} \phi_1(\xi)\, d\xi. \tag{7.4.7}$$

Finally for the one-dimensional equations we examine the inhomogeneous wave equation

$$\frac{\partial^2 \Phi}{\partial t^2} = c^2 \frac{\partial^2 \Phi}{\partial x^2} - \rho(x, t), \tag{7.4.8}$$

with the same initial conditions as in the previous example. In this case we assume that the double integral

$$\iint \rho(x, t)\, dx\, dt,$$

taken over every finite sub-domain of $t > 0$, $-\infty < x < \infty$, exists.

We convert to the characteristic coordinates (ξ, η) introduced in (7.2.5). The equation then has the form

$$c^2 \frac{\partial^2 \Phi}{\partial \xi\, \partial \eta} = \rho\left(\xi + \eta, \frac{\eta - \xi}{c}\right). \tag{7.4.9}$$

On $t = 0$, $\xi = \eta = x/2$, and so the first boundary condition transforms to $\Phi(\xi, \eta) = \phi_0(2\xi)$ on the line $\xi - \eta = 0$. To find the equivalent of the second condition we notice that on $t = 0$,

$$\frac{\partial \Phi}{\partial t} = \phi_1(x) \quad \text{and} \quad \frac{\partial \Phi}{\partial x} = \phi_0'(x),$$

and so

$$\frac{\partial \Phi}{\partial \xi} = \frac{\partial \Phi}{\partial x}\frac{\partial x}{\partial \xi} + \frac{\partial \Phi}{\partial t}\frac{\partial t}{\partial \xi} = \phi_0'(2\xi) - c^{-1}\phi_1(2\xi),$$

on $\xi - \eta = 0$.

If we integrate (7.4.9) with respect to η we obtain

$$c^2 \frac{\partial \Phi}{\partial \xi} = F(\xi) + \int^\eta \rho\left(\xi + \sigma, \frac{\sigma - \xi}{2}\right) d\sigma,$$

where $F(\xi)$ is an arbitrary function of ξ. When we use the second boundary condition, this reduces to

$$c^2\frac{\partial \Phi}{\partial \xi} = c^2\phi_0'(2\xi) - c\phi_1(2\xi) + \int_\xi^\eta \rho\left(\xi + \sigma, \frac{\sigma - \xi}{c}\right) d\sigma. \tag{7.4.10}$$

If we now integrate with respect to ξ we see that

$$2c^2\Phi(\xi, \eta) = G(\eta) + c^2\phi_0(2\xi) - c\int^{2\xi} \phi_1(\tau)\, d\tau + 2\int^\xi d\tau \int_\tau^\eta \rho\left(\tau + \sigma, \frac{\sigma - \tau}{c}\right) d\sigma,$$

where $G(\eta)$ is an arbitrary function of η. When we use the first boundary condition this is reduced to:

$$2c\Phi(\xi, \eta) = c\{\phi_0(2\eta) + \phi_0(2\xi)\} + \int_{2\xi}^{2\eta} \phi_1(\tau)\, d\tau - \frac{2}{c}\int_\xi^\eta d\tau \int_\tau^\eta \rho\left(\tau + \sigma, \frac{\sigma - \tau}{c}\right) d\sigma. \tag{7.4.11}$$

If the double integral at the end of (7.4.11) is excluded, it will be seen that the residual terms are just the solution (7.4.7) found above. It is convenient in the double integral to perform a new transformation of the variables (see Figs. 7.6 and 7.7). We put

$$2\tau = x' - ct' \quad \text{and} \quad 2\sigma = x' + ct', \tag{7.4.12}$$

so that

$$d\sigma\, d\tau = \frac{\partial(\tau, \sigma)}{\partial(x', t')}\, dx'\, dt' = \begin{vmatrix} c/2 & \frac{1}{2} \\ -c/2 & \frac{1}{2} \end{vmatrix} dx'\, dt' = \frac{c}{2}\, dx'\, dt'. \tag{7.4.13}$$

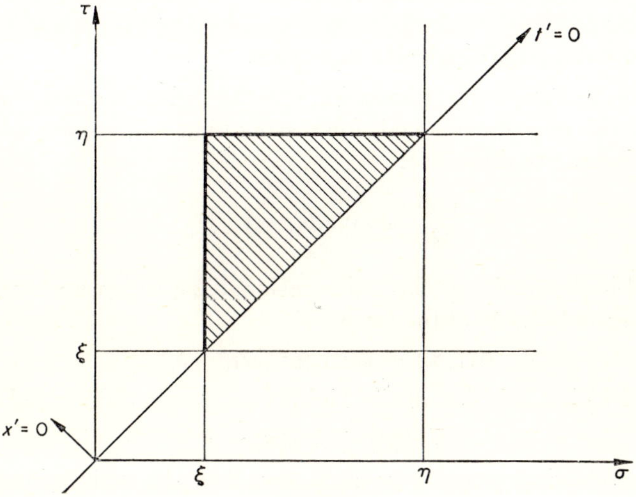

Fig. 7.6

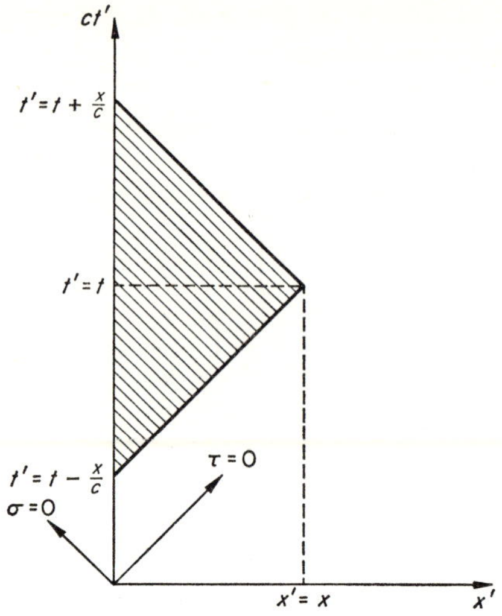

Fig. 7.7

Hence

$$\frac{2}{c}\int_{\xi}^{\eta} d\tau \int_{\tau}^{\eta} \rho\left(\tau + \sigma, \frac{\sigma - \tau}{c}\right) d\sigma = \int_{0}^{t} dt' \int_{x-c(t-t')}^{x+c(t-t')} \rho(x', t')\, dx'. \tag{7.4.14}$$

An alternative form of this double integral can be obtained by writing $t - t'$ instead of t'; it is

$$\int_{0}^{t} dt' \int_{x-ct'}^{x+ct'} \rho(x', t - t')\, dx'. \tag{7.4.15}$$

7.5 The related Helmholtz equation

It is most convenient to examine the three-dimensional wave equation after the one-dimensional, for reasons that will appear later in the chapter. We write therefore:

$$\frac{\partial^2 \Phi}{\partial t^2} = c^2 \nabla^2 \Phi(t, \mathbf{x}) \tag{7.5.1}$$

where ∇^2 is the ordinary three-dimensional Laplacian operator. There are solutions of (7.5.1) of the form

$$\Phi(t, \mathbf{x}) = \phi(\mathbf{x}) \cos (c\lambda t + \delta_\lambda) \tag{7.5.2}$$

wherein $\phi(\mathbf{x})$ is a solution of the Helmholtz partial differential equation

$$\nabla^2 \phi(\mathbf{x}) + \lambda^2 \phi(\mathbf{x}) = 0. \tag{7.5.3}$$

In the following sections we shall, therefore, be mainly interested in finding solutions of equation (7.5.3).

It is of interest to notice that we can generalize the uniqueness proof given in Section 6.3 to apply also to the equation

$$\nabla^2\phi(\mathbf{x}) - \lambda^2\phi(\mathbf{x}) = 0.$$

Let $\phi_1(\mathbf{x})$ and $\phi_2(\mathbf{x})$ be two solutions of this equation, both satisfying the same boundary conditions, that either the function or its normal derivative takes a given value at each point of a closed boundary. Then if, as before, $\phi_3(\mathbf{x}) = \phi_1(\mathbf{x}) - \phi_2(\mathbf{x})$ we shall find, following the argument of Section 6.3, that

$$0 = \int_V \phi_3(\mathbf{x})\{\nabla^2\phi_3 - \lambda^2\phi_3\}\, d\tau = -\int_V [\{\text{grad}\, \phi_3\}^2 + \lambda^2\{\phi_3(\mathbf{x})\}^2]\, d\tau.$$

Since this implies that both grad $\phi_3(\mathbf{x})$ and $\phi_3(\mathbf{x})$ vanish identically at each point of V, we need not add the rider that under Neumann boundary conditions $\phi_3(\mathbf{x})$ could be constant, which applied to the Laplace equation.

This proof cannot be extended to (7.5.3) as the two terms in the second integral above would occur with opposite sign. We may easily see that there is no uniqueness theorem for (7.5.3), as both

$$\phi(\mathbf{x}) = 0 \quad \text{and} \quad \phi(\mathbf{x}) = \sin mx \sin ny \sin pz,$$

with m, n and p, all integers, satisfy

$$\nabla^2\phi(\mathbf{x}) + (m^2 + n^2 + p^2)\phi(\mathbf{x}) = 0,$$

and vanish on the planes $x = 0$, $x = \pi$, $y = 0$, $y = \pi$, $z = 0$ and $z = \pi$. It is in fact easy to see that this non-uniqueness is essential if we are to be able to derive appropriate eigenvalue problems.

7.6 Waves in a finite rectangular cavity

We may clearly look for separable solutions of (7.5.3). If we choose Cartesian coordinates we may proceed as in Section 6.8. Put

$$\phi(x, y, z) = X(x)\, Y(y)\, Z(z),$$

and then

$$\frac{X''(x)}{X(x)} + \frac{Y''(y)}{Y(y)} + \frac{Z''(z)}{Z(z)} + \lambda^2 = 0. \tag{7.6.1}$$

Typical boundary conditions would be that on a closed boundary, in this case a rectangular parallelepiped, either $\phi(\mathbf{x})$ or its normal derivative would be given for all time. These must be taken in conjunction with two initial conditions which, for the original equation, (7.5.1), might take the form that at $t = 0$, $\Phi(0, x, y, z)$ and $\partial\Phi/\partial t$ would both be given.

Let us, for example, suppose that the boundary conditions are that $\Phi(t, x, y, z) = 0$, and therefore $\phi(x, y, z) = 0$, on $x = 0$, $x = a$,

$y = 0$, $y = b$, $z = 0$ and $z = c$. Then, as we have seen above, we may satisfy this by

$$\phi(x, y, z) = A_{mnp} \sin\left(\frac{m\pi x}{a}\right) \sin\left(\frac{n\pi y}{b}\right) \sin\left(\frac{p\pi z}{c}\right), \qquad (7.6.2)$$

provided

$$\lambda^2 = \pi^2\left(\frac{m^2}{a^2} + \frac{n^2}{b^2} + \frac{p^2}{c^2}\right)^{1/2}.$$

Thus we have a possible solution of (7.5.1):

$$\Phi(t, x, y, z) = \sum_{n=1}^{\infty} \sum_{m=1}^{\infty} \sum_{p=1}^{\infty} A_{mnp}(t) \sin\left(\frac{m\pi x}{a}\right) \sin\left(\frac{n\pi y}{b}\right) \sin\left(\frac{p\pi z}{c}\right), \qquad (7.6.3)$$

where

$$A_{mnp}(t) = a_{mnp} \cos\left\{\pi ct\left(\frac{m^2}{a^2} + \frac{n^2}{b^2} + \frac{p^2}{c^2}\right)^{1/2}\right\} + b_{mnp} \sin\left\{\pi ct\left(\frac{m^2}{a^2} + \frac{n^2}{b^2} + \frac{p^2}{c^2}\right)^{1/2}\right\}.$$

The two sets (a_{mnp}) and $\{b_{mnp}\}$ are constants which can be identified if the two initial functions $\Phi(0, x, y, z)$ and $(\partial\Phi/\partial t)_{t=0}$ are expanded in a Fourier treble sine series.

7.7 Waves in a finite spherical cavity

If instead of Cartesian coordinates in (7.5.3) we use spherical polar coordinates, results are obtained which are sufficiently different from the equivalent results for the Laplace equation to be worth noting in detail. If we substitute

$$\phi(R, \theta, \psi) = \rho(R) S_\nu(\theta, \psi), \qquad (7.7.1)$$

where $S_\nu(\theta, \psi)$ is a spherical harmonic of order ν, into (7.5.3), the resulting differential equation for $\rho(R)$ is

$$\frac{d}{dR}\left(R^2 \frac{d\rho}{dR}\right)\{\lambda^2 R^2 - \nu(\nu + 1)\}\rho(R) = 0, \qquad (7.7.2)$$

or

$$R^2 \frac{d\rho}{dR} + 2R\frac{d\rho}{dR} + \{\lambda^2 R^2 - \nu(\nu + 1)\}\rho(R) = 0. \qquad (7.7.3)$$

If we make the substitution

$$\rho(R) = R^\alpha \eta(R),$$

this becomes

$$R^2 \frac{d^2\eta}{dR^2} + 2(\alpha + 1)R\frac{d\eta}{dR} + \{\lambda^2 R^2 + (\alpha - \nu)(\alpha + \nu + 1)\}\eta(R) = 0, \qquad (7.7.4)$$

and with the choice

$$2\alpha + 1 = 0,$$

reduces to the Bessel equation of order $\nu + \frac{1}{2}$

$$R^2\frac{d^2\eta}{dR^2} + R\frac{d\eta}{dR} + \{\lambda^2R^2 - (\nu + \tfrac{1}{2})^2\}\eta(R) = 0. \tag{7.7.5}$$

We expect to be able to apply the same arguments used in the previous chapter to establish that ν is an integer n. In this case the solution of (7.7.3) is

$$\rho(R) = R^{-1/2}\{AJ_{n+1/2}(\lambda R) + BJ_{-n-1/2}(\lambda R)\}. \tag{7.7.6}$$

7.8 A general solution in three dimensions

We might of course interest ourselves in solutions with spherical symmetry. This would be equivalent to making the substitution $\nu = 0$ into (7.7.1) and (7.7.2). Instead, however, we revert to the original wave equation (7.5.1) which becomes

$$\frac{\partial^2\Phi(t, R)}{\partial t^2} = \frac{c^2}{R^2}\frac{\partial}{\partial R}\left\{R^2\frac{\partial\Phi(t, R)}{\partial R}\right\}. \tag{7.8.1}$$

This may be rewritten

$$R\frac{\partial^2\Phi(t, R)}{\partial t^2} = c^2\frac{\partial^2}{\partial R^2}\{R\Phi(t, r)\} \tag{7.8.2}$$

from which it is clear that $R\Phi(t, R)$ satisfies the one-dimensional wave equation (7.2.1), with R written for x. Hence a solution of (7.8.1) is

$$\Phi(t, R) = R^{-1}\{f(R - ct) + g(R + ct)\}, \tag{7.8.3}$$

where as before $f(R)$ and $g(R)$ are arbitrary functions.

We may use this solution to obtain further information about more general solutions of (7.5.1). The equation (7.5.1) is invariant under a change of origin, and so $R^{-1}f(R - ct)$ remains a solution of (7.5.1) when we write

$$R^2 = (x - x')^2 + (y - y')^2 + (z - z')^2, \tag{7.8.4}$$

where $\mathbf{x}' = (x', y', z')$ is an arbitrary point. Since, moreover, solutions of the equation are additive, if we have a finite set of points $(\mathbf{x}_k')$ and put

$$R_k^2 = (x - x_k')^2 + (y - y_k')^2 + (z - z_k')^2,$$

then for a set of arbitrary constants (A_k):

$$\sum_k A_kR_k^{-1}f(R_k - ct)$$

is also a solution. Provided that the convergence is sufficient we can extend this to an infinite set of points. Finally, if we can justify double differentiation under the integral sign, we see that a solution is

$$\Phi(t, x, y, z) = \int_{V'} F(x', y', z')R^{-1}f(R - ct)\,dx'\,dy'\,dz', \qquad (7.8.5)$$

where $F(x', y', z')$ is an arbitrary function (other than the limitations imposed by the differentiation requirements), and the integral is to be taken over some connected three-dimensional set of points V'. An alternative form for (7.8.5) can be obtained by the transformation

$$x' = x + R\sin\theta\cos\psi,$$
$$y' = y + R\sin\theta\sin\psi,$$

and

$$z' = z + R\cos\theta,$$

when it becomes

$$\Phi(t, x, y, z) = \int_{V'} F(x, + R\sin\theta\cos\psi, y + R\sin\theta\sin\psi, z + R\cos\theta) \times f(R - ct)\,R\,dR\sin\theta\,d\theta\,d\psi. \qquad (7.8.6)$$

We now suppose that $f(R)$ is defined for all positive and negative values of R, has a continuous second derivative,

$$f(R) = 0 \quad \text{if} \quad |R| > \varepsilon$$

and

$$\int_{-\varepsilon}^{\varepsilon} f(R)\,dR = 1.$$

We further suppose that V' contains the whole of the spherical shell $ct - \varepsilon \leqslant R \leqslant ct + \varepsilon$. Then by the mean value theorem:

$$\Phi(t, x, y, z) = ct\int_0^{\pi} d\theta\sin\theta\int_{-\pi}^{\pi} F(x + \bar{R}\sin\theta\cos\psi, y + \bar{R}\sin\theta\sin\psi, z + \bar{R}\cos\theta)\,d\psi$$

where $ct - \varepsilon < \bar{R} < ct + \varepsilon$. In the limit therefore as $\varepsilon \to 0$ we have a solution

$$\Phi(t, x, y, z) = 4\pi ct M_t\{F(\mathbf{x}')\} \qquad (7.8.7)$$

where $M_t\{F(\mathbf{x}')\}$ is the average value of $F(\mathbf{x}')$ taken over the surface of a sphere of centre $\mathbf{x} = (x, y, z)$ and radius ct. The restrictions on $F(\mathbf{x}')$ necessary for this will ensure that $M_0\{F(\mathbf{x}')\}$ is finite and so this solution vanishes when $t = 0$. If we differentiate (7.8.7) and put $t = 0$, we see that

$$\left\{\frac{\partial\Phi}{\partial t}\right\}_{t=0} = 4\pi c M_0\{F(\mathbf{x}')\} = 4\pi c F(x, y, z). \qquad (7.8.8)$$

Thus we see that the solution of (7.5.1) which satisfies the initial conditions that

$$\Phi(0, x, y, z) = 0 \quad \text{and} \quad \left\{\frac{\partial}{\partial t}\Phi(t, x, y, z)\right\}_{t=0} = \phi_1(x, y, z)$$

is

$$\Phi(t, x, y, z) = tM_t\{\phi_1(x', y', z')\}. \tag{7.8.9}$$

If $\Phi(t, x, y, z)$ is a solution of (7.5.1) then $\partial\Phi/\partial t$ will also be a solution, provided that the partial differentiations commute, as we can see by differentiating (7.5.1) once with respect to time. Hence another solution of (7.5.1) is

$$4\pi c\frac{\partial}{\partial t}[tM_t\{F(\mathbf{x}')\}]. \tag{7.8.10}$$

This solution takes the value

$$4\pi cF(\mathbf{x})$$

when $t = 0$. Its first derivative with respect to time is

$$4\pi c\frac{\partial^2}{\partial t^2}[tM_t\{F(\mathbf{x}')\}],$$

and since the expression contained within the square brackets is a solution of the wave equation, it can be written alternatively as

$$4\pi c^3 t\nabla^2 Mt\{F(\mathbf{x}')\}.$$

It will therefore follow that for this solution $\partial\Phi/\partial t = 0$ when $t = 0$. Thus the solution of (7.5.1) satisfying the initial conditions

$$\Phi(0, x, y, z) = \phi_0(x, y, z), \quad \text{and} \quad \left\{\frac{\partial}{\partial t}\Phi(t, x, y, z)\right\}_{t=0} = 0$$

is

$$\Phi(t, x, y, z) = \frac{\partial}{\partial t}[tM_t\{\phi_0(x, y, z)\}]. \tag{7.8.11}$$

Finally therefore combining (7.8.9) and (7.8.11), the solution satisfying the initial conditions

$$\Phi(0, x, y, z) = \phi_0(x, y, z) \quad \text{and} \quad \left\{\frac{\partial}{\partial t}\Phi(t, x, y, z)\right\} = \phi_1(x, y, z),$$

is

$$\Phi(t, x, y, z) = \frac{\partial}{\partial t}[tM_t\{\phi_0(x, y, z)\}] + tM_t\{\phi_1(x, y, z)\}. \tag{7.8.12}$$

We see therefore that the solution of the equation at a point $\mathbf{x}$ and at a time t is determined solely by the values of the initial data at points lying a distance ct from $\mathbf{x}$, and by the rate of change of the initial values measured at points distance ct, and along the radius vector directed away from **x**. Conversely, the particular value taken by the solution at a point will influence the solution, at a time t later, only at points which lie at a distance ct away. This is equivalent to a statement of Huyghens' principle, frequently expressed by the statement that given a propagating wave front at any instant, the equivalent front at a time t later is the envelope of all spheres of radius ct whose centres lie on the initial wave front.

We therefore see that, as in one dimension, waves satisfying the three-dimensional wave equation exhibit Huyghens' principle.

7.9 The corresponding solution in two dimensions

Having considered the three-dimensional wave equation in the previous sections, we may now obtain results for the two-dimensional wave equation which are special cases of the earlier work. This is called using the method of dimension reduction. It is of frequent use in studying the wave equation.

Let us consider the result expressed by equation (7.8.9). Instead of using spherical polar coordinates as above, however, we shall now use cylindrical polars. Hence we shall put

$$x' = x + r\cos\theta,\ y' = y + r\sin\theta,\ z' = z + \zeta.$$

The integral is over the surface of the sphere

$$r^2 + \zeta^2 = c^2t^2. \tag{7.9.1}$$

On this sphere, an element of surface area is

$$r\,d\theta\{(dr)^2 + (d\zeta)^2\}^{1/2}, \tag{7.9.2}$$

but since from (7.9.1)

$$r\,dr + \zeta\,d\zeta = 0,$$

(7.8.9) may be written in full as

$$\begin{aligned}\Phi(t, x, y, z) = \frac{1}{4\pi c}\int_0^{ct}\frac{dr\,r}{(c^2t^2 - r^2)^{1/2}} &\times \int_{-\pi}^{\pi}[\phi_1\{x + r\cos\theta, y + r\sin\theta, z + (c^2t^2 - r^2)^{1/2}\}\\ &+ \phi_1\{x + r\cos\theta, y + r\sin\theta, z - (c^2t^2 - r^2)^{1/2}\}]\,d\theta.\end{aligned} \tag{7.9.3}$$

If the problem is two-dimensional, we may assume that ϕ_1 is independent of z. Thus we may rewrite (7.9.3) as

$$\Phi(t, x, y) = \frac{1}{c}\int_0^{ct}\phi_1^*(x, y, r)(c^2t^2 - r^2)^{-1/2}r\,dr, \tag{7.9.4}$$

where

$$\phi_1^*(x, y, r) = \frac{1}{2\pi}\int_{-\pi}^{\pi}\phi_1(x + r\cos\theta, y + r\sin\theta)\,d\theta,$$

and is the mean value of $\phi_1(x', y')$ taken round a circle with centre (x, y) and radius r.

It now appears from (7.9.4) that the solution of the equation at time t, at a given point (x, y), depends on values taken by the initial functions $\phi_1(x', y')$ at every point within a circle of radius ct and centre (x, y). This is not in accord with the Huyghens property, and so the principle is not applicable to two-dimensional problems.

EXERCISES

7.1. Solve the one-dimensional wave equation

$$\frac{\partial^2 \Phi}{\partial t^2} = c^2 \frac{\partial^2 \Phi}{\partial x^2}$$

with the (consistent) initial conditions that on $t = 0$, $\Phi(x, 0) = \phi_0(x)$, $\partial\Phi/\partial t = \phi_1(x)$, and with the boundary conditions $\Phi(0, t) = 0$, $\partial\Phi/\partial x = 0$ at $x = l$, for all $t > 0$. (Finite string fixed at $x = 0$, and attached at $x = l$ to a ring free to slide on a vertical frictionless wire.)

7.2. Find the solution of the wave equation

$$\frac{\partial^2 \Phi}{\partial t^2} = c^2 \frac{\partial^2 \Phi}{\partial x^2} + f(x, t)$$

in $t > 0$, $-\infty < x < \infty$, if on $t = 0$ both Φ and $\partial\Phi/\partial t$ vanish and $f(x, t) = \exp(-x^2)$, $0 \leqslant t < 1$ and $f(x, t) = 0$, $t > 1$.

7.3. Find the solution outside the sphere $R = a$, of the spherically-symmetric wave equation, if on $R = a$, $\Phi(a, t) = \sin \omega t$, for all $t > 0$, and $\Phi \to 0$ as $R \to \infty$.

7.4. Transform the equation

$$\frac{\partial^2 \Phi}{\partial t^2} = c^2 \nabla^2 \Phi$$

by the substitution $r^2 = x^2 + y^2 = -4\xi\eta$, $z = -(\xi + \eta)$ and hence show that a solution can be written as

$$\Phi = \exp\{i\lambda(\xi + \eta \pm ct\} L_\nu(2ik\xi) L_\nu(2ik\eta)$$

where $L_\nu(\zeta)$ satisfies the Laguerre equation

$$\zeta W'' + (1 - \zeta)W' + \nu W = 0.$$

7.5. If $x = \cosh \xi \cos \eta$, $y = \sinh \xi \sin \eta$, show that the two-dimensional Helmholtz equation

$$\frac{\partial^2 \Phi}{\partial x^2} + \frac{\partial^2 \Phi}{\partial y^2} + \lambda^2 \Phi = 0$$

reduces to

$$\frac{\partial^2 \Phi}{\partial \xi^2} + \frac{\partial^2 \Phi}{\partial \eta^2} + \lambda^2(\cosh^2 \xi - \cos^2 \eta) = 0,$$

and so express the solution in Mathieu functions (see [24]).

ANSWERS

7.1. $$\Phi = \frac{2}{l}\sum_{n=0}^{\infty}\left\{\cos k_n ct \int_0^l \phi_0(\xi)\sin k_n\xi\,d\xi + c^{-1}\sin k_n ct\int_0^l \phi_1(\xi)\sin k_n\xi\,d\xi\right\}\sin k_n x$$

where $$2lk_n = (2n+1)\pi.$$

7.2. $$2c\Phi(x,t) = tF(x,t) - (t-1)F(x,t-1) - c\int_{t-1}^{t} t'[\exp\{-(x+ct')^2\} - \exp\{-(x-ct')^2\}]dt'$$

where $$F(x,t) = \int_{x-ct}^{x+ct}\exp(-\xi^2)\,d\xi.$$

7.3. $$\Phi = \frac{a}{R}\sin\left\{\frac{\omega}{c}(ct - R + a)\right\}.$$

CHAPTER 8

The Diffusion or Heat-conduction Equation

8.1 The physical significance

This equation, which we shall take in the form

$$\frac{\partial \Phi}{\partial t} = k\nabla^2\Phi, \ (k > 0), \tag{8.1.1}$$

is parabolic, and we shall use it to typify the behaviour of such types of linear partial differential equations.

It arises in mathematical physics from a combination of a continuity equation like (7.1.3) with a second equation

$$\mathbf{U} = -k \operatorname{grad} \phi, \tag{8.1.2}$$

which makes the rate of transport proportional to the gradient of the quantity ϕ, and directed from large ϕ to small ϕ. This is physically plausible for the diffusion of molecules or ions. Since, on the kinetic theory, temperature is related to the mean velocity of molecules, the process of heat conduction just represents the diffusion of molecules of greater average speed among those of lower speed.

All linear, time-dependent problems of mathematical physics are basically wave problems or diffusion problems. In the former case effects propagate at a finite speed, and a disturbance at a given point at one instant of time does not influence another point until a definite later instant, when its effect is felt to the full. In a diffusion problem a disturbance at a given point at one instant effects instantaneously all other points, but with vanishingly small effect, and the effect will not rise above an arbitrarily given threshold level until a finite time later. This time will increase with the distance from the initial point.

8.2 The finite bar

As usual we shall start with the one-dimensional equation

$$\frac{\partial \Phi}{\partial t} = k\frac{\partial^2 \Phi}{\partial x^2}. \tag{8.2.1}$$

If we seek a separable solution of the form

$$\Phi(x, t) = \tau(t)X(x),$$

we shall derive

$$\frac{\tau'(t)}{\tau(t)} = k\frac{X''(x)}{X(x)}. \tag{8.2.2}$$

Hence a solution is

$$\Phi(x, t) = \sum_{\lambda} A_\lambda \exp(-k\lambda^2 t) \cos(\lambda x + \varepsilon_\lambda). \tag{8.2.3}$$

We need two boundary conditions and one initial condition to solve equation (8.2.1). If the boundary conditions are those for a finite bar, this will naturally give rise to an eigenvalue problem for which an expansion like (8.2.3) would be suitable.

Suppose for example that the boundary conditions are that at $x = 0$, $\Phi(x, t) = \alpha$ and at $x = l$, $\Phi(x, t) = \beta$ for all $t > 0$, where α and β are given constants. Then if we put

$$\Phi(x, t) = \alpha + (\beta - \alpha)x + \phi(x, t), \tag{8.2.4}$$

$\phi(x, t)$ is also a solution of equation (8.2.1), and satisfies the boundary conditions $\phi(0, t) = \phi(l, t) = 0$, $t > 0$. For $\phi(x, t)$, from (8.2.3), a solution satisfying the boundary conditions is:

$$\phi(x, t) = \sum_{n=1}^{\infty} A_n \exp(-kn^2\pi^2 t/l^2) \sin(n\pi x/l). \tag{8.2.5}$$

If a suitable initial distribution $\Phi(x, 0) = \Phi_0(x)$, $0 \leqslant x \leqslant l$ is given, then the coefficients A_n may be found by comparing (8.2.5) at $t = 0$, term by term with the Fourier sine series for $\Phi_0(x) - \alpha - (\beta - \alpha)x$. It is worth-while noting that since the time functions in (8.2.5) are exponentially decreasing, and since n^2 appears in their argument, the equation could, for most practical purposes, be approximated, after a short initial period, by its leading term

$$A_1 \exp(-k\pi^2 t/l^2) \sin(\pi x/l). \tag{8.2.6}$$

8.3 The semi-infinite bar

When the x-interval is semi-infinite or infinite the most convenient method of solution is by an integral transform. The generalization of (8.2.5) suggests a Fourier sine transform (see Appendix 3). The advantage is that we can solve some typical problems directly. In a later section we shall consider another method of finding solutions, but this suffers from the disadvantage of being primarily an *ad hoc* method.

We consider three problems in this section.

(*a*) We seek a solution of (8.2.1) satisfying the boundary conditions $\Phi(0, t) = \alpha$ and $\Phi(x, t) \to \beta$ as $x \to \infty$, for all $t > 0$, and the initial condition that $\Phi(x, 0) = \beta$, $x > 0$.

We put

$$\Phi(x, t) = \beta + \phi(x, t), \tag{8.3.1}$$

and

$$\bar{\phi}(\lambda, t) = \sqrt{\frac{2}{\pi}} \int_0^\infty \phi(x, t) \sin \lambda x \, dx. \tag{8.3.2}$$

Then if we make the necessary substitutions into (8.2.1) we see that $\bar{\phi}(t, \lambda)$ satisfies the equation

$$\frac{d\bar{\phi}}{dt} = -k\left\{\lambda^2\bar{\phi}(\lambda, t) - \sqrt{\frac{2}{\pi}}\lambda(\alpha - \beta)\right\}, \tag{8.3.3}$$

with the initial condition that $\bar{\phi}(\lambda, 0) = 0$. Hence

$$\bar{\phi}(\lambda, t) = \sqrt{\frac{2}{\pi}} \frac{(\alpha - \beta)}{\lambda}\{1 - \exp(-k\lambda^2 t)\}. \tag{8.3.4}$$

The solution can then be obtained from the inversion integral

$$\begin{aligned} \phi(x, t) &= \sqrt{\frac{2}{\pi}} \int_0^\infty \bar{\phi}(\lambda, t) \sin \lambda x \, d\lambda \\ &= \frac{2(\alpha - \beta)}{\pi} \int_0^\infty \frac{\{1 - \exp(-k\lambda^2 t)\} \sin \lambda x \, d\lambda}{\lambda}. \end{aligned} \tag{8.3.5}$$

This can be simplified. First we can use the well-known result that if $x > 0$

$$\int_0^\infty \frac{\sin \lambda x}{\lambda} d\lambda = \frac{\pi}{2},$$

and secondly we have

$$\begin{aligned} \int_0^\infty \frac{\exp(-k\lambda^2 t) \sin \lambda x \, d\lambda}{\lambda} &= \int_0^x d\xi \int_0^\infty \exp(-k\lambda^2 t) \cos \lambda\xi \, d\lambda, \\ &= \frac{1}{2}\sqrt{\left(\frac{\pi}{kt}\right)} \int_0^x d\xi \exp(-\xi^2/4kt), \end{aligned}$$

if we use another well-known integral. If then we write

$$U(x, t) = \frac{2}{\sqrt{\pi}} \int_{x/2\sqrt{(kt)}}^\infty \exp(-\eta^2) \, d\eta, \tag{8.3.6}$$

the solution of problem (*a*) is:

$$\Phi(x, t) = \beta + (\alpha - \beta)U(x, t). \tag{8.3.7}$$

(*b*) In the second problem the only change we shall make in the conditions imposed on the solution in (*a*) is to assume that $\alpha = \alpha(t)$ is a function of time.

Clearly, as before, we may derive an equation like (8.3.3), and so we must insist that $\alpha(t)$ is well enough behaved for this to be soluble. The corresponding solution is now:

$$\bar{\phi}(\lambda, t) = \sqrt{\frac{2}{\pi}}k\lambda \exp(-k\lambda^2 t)\int_0^t \{\alpha(\sigma) - \beta\} \exp(k\lambda^2\sigma)\, d\sigma,$$

$$= \sqrt{\frac{2}{\pi}}k\lambda\int_0^t \alpha(\sigma) \exp\{-k\lambda^2(t-\sigma)\}\, d\sigma$$

$$- \sqrt{\frac{2}{\pi}}\frac{\beta}{\lambda}\{1 - \exp(-k\lambda^2 t)\},$$

$$= \sqrt{\frac{2}{\pi}}k\lambda\int_0^t \alpha(t-\tau) \exp(-k\lambda^2\tau)\, d\tau$$

$$- \sqrt{\frac{2}{\pi}}\frac{\beta}{\lambda}\{1 - \exp(-k\lambda^2 t)\}. \tag{8.3.8}$$

Now

$$\int_0^\infty \lambda \exp(-k\lambda^2\tau) \sin \lambda x\, d\lambda = -\frac{d}{dx}\int_0^\infty \exp(-k\lambda^2\tau) \cos \lambda x\, d\lambda$$

$$= \frac{x\sqrt{\pi}}{4(k\tau)^{3/2}} \exp(-x^2/4k\tau).$$

Hence

$$\phi(x, t) = \frac{x}{2(2k)^{1/2}}\int_0^t \alpha(t-\tau)\tau^{-3/2} \exp\{-x^2/4k\tau\}\, d\tau - \beta U(x, t). \tag{8.3.9}$$

(*c*) In the final problem, we keep the same boundary conditions as in problem (*a*) but replace the initial condition by

$$\Phi(x, 0) = B(x) + \beta, \quad x > 0,$$

where we also assume that $B(x) \to 0$ as $x \to \infty$.

As before we use the substitution of (8.3.1) and derive the equation (8.3.3). The initial condition is however replaced by

$$\phi(\lambda, 0) = \sqrt{\frac{2}{\pi}}\int_0^\infty B(x) \sin \lambda x\, dx = \bar{B}(\lambda). \tag{8.3.10}$$

This will converge as an infinite integral, provided $B(x)$ is absolutely integrable on every finite interval.

The new initial condition will add to the right-hand side of (8.3.4) a term

$$\bar{B}(\lambda) \exp(-k\lambda^2 t). \tag{8.3.11}$$

This term may be interpreted quite simply if we use the convolution theorem for the Fourier transform [15].

Since we are using sine transforms, this is equivalent to the assumption that the functions involved are odd. Hence the convolution theorem takes the form that if $\bar{f}(\lambda)$ is the sine transform of $f(x)$ and $\bar{g}(\lambda)$ of $g(x)$ then $\lambda\bar{f}(\lambda)\bar{g}(\lambda)$ is the transform of:

$$(2\pi)^{1/2}\int_0^\infty f(\xi)\{g'(\,|\,x-\xi\,|\,)-g'(x+\xi)\}\,d\xi.$$

If we then put

$$\bar{f}(\lambda)=\bar{B}(\lambda),\quad \bar{g}(\lambda)=\lambda^{-1}\exp\,(-k\lambda^2 t),$$

we find that (8.3.11) is the transform of

$$\tfrac{1}{2}(kt)^{-1/2}\int_0^\infty B(\xi)[\exp\,\{-(x-\xi)^2/4kt\}-\exp\{-(x+\xi)^2/4kt\}]\,d\xi. \tag{8.3.12}$$

This extra term then completes the solution.

8.4 Similarity solutions

We may try to find solutions of (8.2.1) by another method. Suppose that

$$\Phi(x,t)=t^\rho f(\eta), \tag{8.4.1}$$

where

$$\eta=x(kt)^\nu. \tag{8.4.2}$$

Since $t^{-\rho}\Phi(t,x)$ will then be constant at every point along the trajectories

$$x=C(kt)^{-\nu},$$

where C is a given constant, we shall call a solution of (8.2.1) which has this form a 'similarity' solution. Similarity solutions are frequently of considerable interest in mathematical physics.

If we perform the partial differentiations, and use a prime to denote total differentiation with respect to η, then

$$\frac{\partial\Phi}{\partial t}=\rho t^{\rho-1}f(\eta)+\frac{\nu\eta}{t}t^\rho f'(\eta)=t^{\rho-1}\{\rho f(\eta)+\nu\eta f'(\eta)\},$$

and

$$\frac{\partial^2\Phi}{\partial x^2}=k^{2\nu}t^{\rho+2\nu}f''(\eta).$$

If then $\Phi(t,x)$ is a solution of (8.2.1) it follows that

$$\rho f(\eta)+\nu\eta f'(\eta)=(kt)^{2\nu+1}f''(\eta). \tag{8.4.3}$$

The left-hand side of the equation depends only on η, and so the right-hand side must also depend on η only, and not separately on t or x. Hence

$$2\nu + 1 = 0, \tag{8.4.4}$$

$$\eta = x(kt)^{-1/2},$$

and

$$f''(\eta) + \frac{\eta}{2} f'(\eta) - \rho f(\eta) = 0. \tag{8.4.5}$$

Now differentiate (8.4.5) so that

$$f'''(\eta) + \frac{\eta}{2} f''(\eta) - (\rho - \tfrac{1}{2}) f'(\eta) = 0, \tag{8.4.6}$$

or in other words if $f_n(\eta)$ is a solution of (8.4.5) with the parameter $\rho = n$, then $f_n'(\eta)$ is a solution with the parameter $n - \frac{1}{2}$. If we put $\rho = 0$, we get

$$f_0''(\eta) + \frac{\eta}{2} f_0'(\eta) = 0, \tag{8.4.7}$$

which gives the solution

$$\Phi_0(x, t) = f_0(\eta) = A + B\int_0^{\eta/2} \exp(-\xi^2)\, d\xi, \tag{8.4.8}$$

where A and B are arbitrary constants. The solution we have given in (8.3.7) to the problem (a) is of this type, and the particular values of the constants for this problem could be got from the data.

A second solution is therefore

$$\Phi_{-1/2}(x, t) = t^{-1/2} f_{-1/2}(\eta) = t^{-1/2} f_0'(\eta) = Bt^{-1/2} \exp(-x^2/4kt). \tag{8.4.9}$$

This solution has the properties that for $t > 0$

$$\Phi_{-1/2}(x, t) \to 0 \quad \text{as} \quad x \to \pm\infty$$

and

$$\Phi_{-1/2}(0, t) = Bt^{-1/2}.$$

For all $x \neq 0$, $\Phi_{-1/2}(x, t) \to 0$, both as $t \to 0$ and as $t \to \infty$. Finally we observe that

$$\int_{-\infty}^{\infty} \Phi_{-1/2}(x, t)\, dx = Bt^{-1/2} \int_{-\infty}^{\infty} \exp(-x^2/4kt)\, dx$$

$$= 2Bk^{1/2} \int_{-\infty}^{\infty} \exp(-\xi^2)\, d\xi = 2(\pi k)^{1/2} B.$$

To give a physical interpretation of this solution it is convenient to consider the equation for heat conduction. If the range of x is interpreted as a rod stretching from $-\infty$ to $+\infty$, with constant temperature at $\pm\infty$, and $\Phi(x, t)$ is the temperature increase above this constant value, the integral above is the heat content of the bar, the constant temperature being taken as zero. Thus we have a constant quantity of heat injected into the bar at time $t = 0$, at the point $x = 0$. The bar is then thermally insulated, so that no heat escapes, but the existing content adjusts itself to reach an equilibrium state.

If $B(x)$ is integrable over any finite interval, and is continuous at a given point $x_0 > 0$, then

$$\begin{aligned} t^{-1/2}\int_0^\infty B(\xi) \exp\{-(x_0-\xi)^2/4kt\}\, d\xi &= t^{-1/2}\int_0^{x_0-\varepsilon} B(\xi) \exp\{-(x_0-\xi)^2/4kt\}\, d\xi \\ &+ t^{-1/2}\int_{x_0+\varepsilon}^\infty B(\xi) \exp\{-(x_0-\xi)^2/4kt\}\, d\xi \\ &+ 2B(x_0+h\varepsilon)t^{-1/2}\int_{-\varepsilon}^{\varepsilon} \exp(-\xi^2/4kt)\, d\xi, \end{aligned}$$

where $|h| < 1$ by the mean value theorem, if ε is small enough. Hence

$$\lim_{t\to 0} t^{-1/2}\int_0^\infty B(\xi) \exp\{-(x_0-\xi)^2/4kt\}\, d\xi = 2\sqrt{(k\pi)}B(x_0).$$

Thus

$$\phi(x, t) = \tfrac{1}{2}(k\pi t)^{-1/2}\int_0^\infty B(\xi)[\exp\{-(x-\xi)^2/4kt\} - \exp\{-(x+\xi)^2/4kt\}]\, d\xi,$$

is a solution of (8.2.1) satisfying the boundary conditions that for $t > 0$, $\phi(0, t) = 0$, and $\phi(x, t) \to 0$ as $x \to \infty$, and the initial conditions that at $t = 0$, $\phi(x, 0) = B(x)$.

If we write:

$$G(x, \xi, t) = \tfrac{1}{2}(k\pi t)^{-1/2}[\exp\{-(x-\xi)^2/4kt\} - \exp\{-(x+\xi)^2/4kt\}],$$

then the solution of the problem (*c*) is

$$\Phi(x, t) = \beta + (\alpha - \beta)U(x, t) + \int_0^\infty B(\xi)G(x, \xi, t)\, d\xi. \quad (8.4.10)$$

Since

$$\begin{aligned} \int_0^\infty G(x, \xi, t)\, d\xi &= \tfrac{1}{2}(k\pi t)^{-1/2}\left\{\int_{-x}^\infty \exp\left(\frac{-\eta^2}{4kt}\right) d\eta - \int_x^\infty \exp\left(\frac{-\eta^2}{4kt}\right) d\eta\right\} \\ &= 1 - U(x, t), \end{aligned}$$

we may rewrite (8.4.10) as

$$\Phi(x, t) = \alpha U(x, t) + \int_0^\infty \{B(\xi) + \beta\}G(x, \xi, t)\, d\xi. \quad (8.4.11)$$

Suppose now that at $t = 0$, $\Phi(x, t) = \beta$ for all $x > 0$, and that for $t > 0$, $\Phi(0, t) = \alpha_0$ and $\Phi(x, t) \to \beta$ as $x \to \infty$. The solution is that given in (8.3.7). Hence at a later interval t_1, we have

$$\Phi(x, t_1) = \alpha_0 U(x, t_1) + \beta\{1 - U(x, t_1)\}. \quad (8.4.12)$$

If at this instant α_0 is changed to α_1, we can find the solution for $t > t_1$ from (8.4.11) putting $B(x) = \Phi(x, t_1)$ and replacing t by $t - t_1$. The solution for $t > t_1$ is therefore

$$\Phi(x, t) = \alpha_1 U(x, t - t_1) + \int_0^\infty \Phi(\xi, t_1) G(x, \xi, t - t_1)\, d\xi. \qquad (8.4.13)$$

The change in α has affected only the first term. In particular therefore if $\alpha_1 = \alpha_0$ we could have continued with the solution (8.3.7), and so for $t > t_1$,

$$\alpha_0 U(x, t) + \beta\{1 - U(x, t)\} = \alpha_0 U(x, t - t_1) + \int_0^\infty \Phi(\xi, t_1) G(x, \xi, t - t_1)\, d\xi. \qquad (8.4.14)$$

This may be used to eliminate the final integral from (8.4.13), and so an alternative form for the solution for $t > t_1$ is

$$\Phi(x, t) = \alpha_1 U(x, t - t_1) + \alpha_0\{U(x, t) - U(x, t - t_1)\} + \beta\{1 - U(x, t)\}. \qquad (8.4.15)$$

By an exactly similar argument we may show that if at a later instant $t_2 > t_1$, α_1 is changed discontinuously to α_2, the solution for $t > t_2$ is

$$\Phi(x, t) = \alpha_2 U(x, t - t_2) + \alpha_1\{U(x, t - t_1) - U(x, t - t_2)\} + \alpha_0\{U(x, t) - U(x, t - t_1)\} + \beta\{1 - U(x, t)\}. \qquad (8.4.16)$$

We may clearly generalize this; let $\delta\tau$ be a short, constant interval of time, and let us suppose that at the instant $t = n\,\delta\tau$ the boundary condition is changed discontinuously from

$$\Phi(0, t) = \alpha_{n-1} \quad \text{to} \quad \Phi(0, t) = \alpha_n.$$

Then the appropriate solution in the interval $n\,\delta\tau < t \leqslant (n + 1)\,\delta\tau$ is

$$\Phi(x, t) = \alpha_n U(x, t - n\,\delta\tau) + \sum_{r=0}^{n-1} \alpha_r\{U(x, t - r\,\delta\tau) - U(x, t - \delta\tau - r\,\delta\tau)\} + \beta\{1 - U(x, t)\}. \qquad (8.4.17)$$

Finally we may proceed to the limit where α varies continuously with t. This limiting process can be made quite rigorous provided $\alpha(t)$ is well enough behaved, but since the detail is quite elementary we omit it. We may therefore deduce the solution to problem (*b*) in the form

$$\Phi(x, t) = \beta\{1 - U(x, t)\} - \int_0^t \alpha(\tau)\, dU(x, t - \tau), \qquad (8.4.18)$$

or

$$\Phi(x, t) = \beta\{1 - U(x, t)\} + \int_0^t \alpha(t - \tau)\, dU(x, \tau). \qquad (8.14.19)$$

This can be taken as a statement of the theorem of Duhamel.

8.5 The related Helmholtz equation

If we take the full equation (8.1.1), then

$$\Phi(x, y, z, t) = \phi(x, y, z) \exp(-k\lambda^2 t) \tag{8.5.1}$$

will be a solution of the equation, provided $\phi(x, y, z)$ satisfies the Helmholtz equation

$$\nabla^2\phi + \lambda^2\phi(x, y, z) = 0. \tag{8.5.2}$$

This is the same equation which appeared in our study of the wave equation, and so quite a lot of the theory contained in the previous chapter can be applied, with the obvious modifications, to the diffusion equation. For that reason we shall not consider in detail any other particular problems.

Exercises

8.1. Show that a solution of

$$\frac{\partial \Phi}{\partial t} = \frac{\partial^2 \Phi}{\partial x^2},$$

satisfying the initial conditions $\Phi(x, 0) = 0$, and the boundary conditions

$$\Phi(0, t) = a \cos \omega t,\ \Phi \to 0 \text{ as } x \to \infty$$

is
$$\Phi = ae^{-\sigma t} \cos(\omega t - \sigma x) - \frac{a}{\pi}\int_0^\infty e^{-\xi t} \sin(\xi^{1/2}x)\frac{\xi\, d\xi}{\xi^2 + \omega^2},$$

where $2\sigma^2 = \omega^2$.

8.2. Show that a solution of

$$\frac{\partial \Phi}{\partial t} = \frac{\partial^2 \Phi}{\partial x^2}$$

which vanishes for all x at $t = 0$, which $\to 0$ as $x \to \infty$, so that at $x = 0$

$$\frac{\partial}{\partial t}\{e^{-t}\Phi(0, t)\} + 1 = 0$$

is
$$\frac{2e^t}{\pi}\int_0^t d\tau \int_0^\infty \exp\{-\tau(1 + \xi^2)\}\{\sin \xi x + \xi \cos \xi x\}\frac{\xi\, d\xi}{1 + \xi^2}.$$

8.3. Show that a solution of

$$\frac{\partial \Phi}{\partial t} = \frac{\partial^2 \Phi}{\partial r^2} + \frac{1}{r}\frac{\partial \Phi}{\partial r}$$

in $r > a$, $t > 0$, which vanishes when $t = 0$, and for which $\Phi = \Phi_0$, a constant, on $r = a$, for all $t > 0$ and for which $\Phi \to 0$ as $r \to \infty$ is

$$\Phi(r, t) = \frac{a\Phi_0}{r}\left\{1 - \operatorname{erf}\left(\frac{r - a}{2\sqrt{t}}\right)\right\}$$

$$\left(\operatorname{erf} z = \left(\frac{2}{\pi}\right)^{1/2} \int_0^z \exp(-\zeta^2)\, d\zeta\right).$$

8.4. Show that the equation

$$\frac{\partial \Phi}{\partial y}\frac{\partial^2 \Phi}{\partial x\,\partial y} - \frac{\partial \Phi}{\partial x}\frac{\partial^2 \Phi}{\partial y^2} - \frac{\partial^3 \Phi}{\partial y^3} = 0$$

has a similarity solution $\Phi = x^m f(\eta)$ where $\eta = x^n y$ provided that $m = n + 1$ and then

$$f'''(\eta) = (2n + 1)\{f'(\eta)\}^2 - (n + 1)f(\eta)f''(\eta).$$

CHAPTER 9

Green's Functions

9.1 The simple properties

We return to the point that was made in Section 6.4. Consider again the Dirichlet problem for the Laplace equation (6.1.1). Suppose that the point with position vector $\mathbf{a}$ is interior to the bounded domain V. Let us assume that we can find a function $G(\mathbf{x}, \mathbf{a})$ with the following properties:

(*a*) $G(\mathbf{x}, \mathbf{a})$ satisfies the equation (6.1.1) at every interior point of V, with the exception of $\mathbf{a}$ itself.

(*b*) In the neighbourhood of $\mathbf{x} = \mathbf{a}$, $|\mathbf{x} - \mathbf{a}|^{n-2} G(\mathbf{x}, \mathbf{a})$ is finite, (or when $n = 2$, $G(\mathbf{x}, \mathbf{a})/\log |\mathbf{x} - \mathbf{a}|$ is bounded).

(*c*) On the boundary Σ of V, $G(\mathbf{x}, \mathbf{a}) = 0$,

then we call the function a Green's function for (6.1.1). It will be noticed immediately that conditions (*a*) and (*b*) are satisfied by the fundamental solution of Section 6.4, and it is only condition (*c*) which introduces any new ideas.

Provided that such a function exists, we may again use Green's identity, which we apply to the domain V' between $r = |\mathbf{x} - \mathbf{a}| = \delta$ and Σ.

Then if $\Phi(\mathbf{x})$ is a solution of (6.1.2)

$$\int_{V'} \{G(\mathbf{x}, \mathbf{a}) \nabla^2 \Phi(\mathbf{x}) - \Phi(\mathbf{x}) \nabla^2 G(\mathbf{x}, a)\}\, d\tau = -\Omega_n \int_{V'} \rho(\mathbf{x})\, G(\mathbf{x}, \mathbf{a})\, d\tau, \tag{9.1.1}$$

while, from the identity, the left-hand side is also equal to

$$-\int_\Sigma \Phi(x) \frac{\partial G}{\partial n} dS - \int_{r=\delta} \left\{ G(\mathbf{x}, \mathbf{a}) \frac{\partial \Phi}{\partial r} - \Phi(\mathbf{x}) \frac{\partial G(\mathbf{x}, \mathbf{a})}{\partial r} \right\} dS. \tag{9.1.2}$$

If we now proceed to the limit as $\delta \to 0$ we find, corresponding to (6.4.4), the solution of the Dirichlet problem.

$$(n - 2)\Phi(\mathbf{a}) = \int_V \rho(\mathbf{x})\, G(\mathbf{x}, \mathbf{a})\, d\tau + \Omega_n^{-1} \int_\Sigma \Phi(\mathbf{x}) \frac{\partial G}{\partial n} dS. \tag{9.1.3}$$

If we can solve the single problem of determining the Green's function corresponding to a given boundary, we can then solve all properly posed Dirichlet problems for the same boundary.

9.2 Some elementary Green's functions

If we consider in particular the Laplace equation in two or three dimensions, the Green's function has a simple physical interpretation. It is just the electrostatic potential obtained by placing a unit charge at **a** and earthing the boundary Σ. We may therefore write down the solutions for a number of Dirichlet problems.

(*a*) In two dimensions, suppose we seek a solution of

$$\frac{\partial^2\Phi}{\partial x^2} + \frac{\partial^2\Phi}{\partial y^2} = 0,$$

in the half-plane $x > 0$, which satisfies the boundary conditions $\Phi(x, y) \to 0$ as $x^2 + y^2 \to \infty$, within the half-plane, and $\Phi(0, y)$ a given function of y (with the correct asymptotic behaviour when $|y|$ is large).

Taking $\mathbf{a} = (\alpha, \beta)$, then

$$G(\mathbf{x}, \mathbf{a}) = \tfrac{1}{2}\log[\{(x-\alpha)^2 + (y-\beta)^2\}/\{(x+\alpha)^2 + (y-\beta)^2\}]$$

and so

$$\Phi(\alpha, \beta) = \frac{\alpha}{\pi}\int_{-\infty}^{\infty} \frac{\Phi(0, y)\, dy}{\alpha^2 + (y-\beta)^2}. \tag{9.2.1}$$

(*b*) In two dimensions we seek a solution inside the circle $x^2 + y^2 = \rho^2$, when we know that on the circle

$$\Phi(\rho\cos\theta', \rho\sin\theta') = u(\theta').$$

Then if $|\mathbf{a}| = a$

$$G(\mathbf{x}, \mathbf{a}) = \tfrac{1}{2}\log[a^3\rho\{(x-\alpha)^2 + (y-\beta)^2\}/\{(a^2x - \rho^2\alpha)^2 + (a^2y - \rho^2\beta)^2\}]$$

or if $\alpha = a\cos\theta$, $\beta = a\sin\theta$ and if for the moment we put $x^2 + y^2 = r^2$,

$$G(\mathbf{x}, \mathbf{a}) = \tfrac{1}{2}\log[a\rho\{r^2 - 2ar\cos(\theta - \theta') + a^2\} \times \{a^2r^2 - 2a\rho^2 r\cos(\theta - \theta') + \rho^4\}^{-1}].$$

Hence

$$\Phi(a\cos\theta, a\sin\theta) = \frac{\rho^2 - a^2}{2\pi}\int_0^{2\pi} \frac{u(\theta')\, d\theta'}{\rho^2 - 2a\rho\cos(\theta - \theta') + a^2}, \tag{9.2.2}$$

which with slightly changed notation agrees with (6.7.10).

(*c*) In three dimensions we seek a solution of

$$\frac{\partial^2\Phi}{\partial x^2} + \frac{\partial^2\Phi}{\partial y^2} + \frac{\partial^2\Phi}{\partial z^2} = 0$$

in the half-space $z > 0$, which solution vanishes at infinity, when we know $\Phi(x, y, 0)$.

If we now take $\mathbf{a} = (\alpha, \beta, \gamma)$, then

$$G(\mathbf{x}, \mathbf{a}) = \{(x - \alpha)^2 + (y - \beta)^2 + (z - \gamma)^2\}^{-1/2} - \{(x - \alpha)^2 + (y - \beta)^2 + (z + \gamma)^2\}^{-1/2}$$

and then

$$\Phi(\alpha, \beta, \gamma) = \frac{\gamma}{2\pi}\int_{-\infty}^{\infty}\int_{-\infty}^{\infty} \frac{\Phi(x, y, 0)\, dx\, dy}{\{(x - \alpha)^2 + (y - \beta)^2 + \gamma^2\}^{3/2}} \tag{9.2.3}$$

(*d*) Finally we seek a solution inside the sphere $x^2 + y^2 + z^2 = \rho^2$, and put $\mathbf{x} = (r \sin \theta' \cos \psi', r \sin \theta' \sin \psi', r \cos \theta')$, $\mathbf{a} = (a \sin \theta \cos \psi, a \sin \theta \sin \psi, a \cos \theta)$ and $\mathbf{a}.\mathbf{x} = ar \cos \Theta$, then

$$G(\mathbf{x}, \mathbf{a}) = (r^2 - 2ar \cos \Theta + a^2)^{-1/2} - \frac{\rho}{a}\left(r^2 - 2\frac{\rho^2 r}{a} \cos \Theta + \frac{\rho^4}{a^2}\right)^{-1/2}$$

and

$$\Phi(\alpha, \beta, \gamma) = \frac{\rho(\rho^2 - a^2)}{4\pi}\int_0^{\pi} d\theta' \sin \theta' \int_0^{2\pi} \frac{\Phi(\rho \sin \theta' \cos \psi', \rho \sin \theta' \sin \psi', \rho \cos \theta')\, d\psi'}{(\rho^2 - 2a\rho \cos \Theta + a^2)^{3/2}} \tag{9.2.4}$$

which is in agreement with (6.9.18).

9.3 The property of symmetry

The Green's function for the Laplace equation is symmetric, considered as a function of **x** and **a**. This can be deduced quite simply from the Green's identity.

Let **a** and **b** be two distinct points in the interior of V, and let the domain V'' be obtained from V by removing the interior of the two small spheres $r = |\mathbf{x} - \mathbf{a}| = \delta$ and $r' = |\mathbf{x} - \mathbf{b}| = \delta'$, where it is assumed that δ and δ' are both so small that the whole interior of both spheres is included in V. Then

$$\begin{aligned}\int_{V''} \{G(\mathbf{x}, \mathbf{a})\, \nabla^2 G(\mathbf{x}, \mathbf{b}) - G(\mathbf{x}, \mathbf{b})\, \nabla^2 G(\mathbf{x}, \mathbf{a})\}\, d\tau &= 0, \\ = \int_{r=\delta} \left\{G(\mathbf{x}, \mathbf{b})\frac{\partial}{\partial r} G(\mathbf{x}, \mathbf{a}) - G(\mathbf{x}, \mathbf{a})\frac{\partial}{\partial r} G(\mathbf{x}, \mathbf{b})\right\} dS & \\ + \int_{r'=\delta'} \left\{G(\mathbf{x}, \mathbf{b})\frac{\partial}{\partial r'} G(\mathbf{x}, \mathbf{a}) - G(\mathbf{x}, \mathbf{a})\frac{\partial}{\partial r'} G(\mathbf{x}, \mathbf{b})\right\} dS.&\end{aligned} \tag{9.3.1}$$

In the limit as $\delta \to 0$ and $\delta' \to 0$, then

$$0 = (n - 2)\Omega_n\{G(\mathbf{b}, \mathbf{a}) - G(\mathbf{a}, \mathbf{b})\}. \tag{9.3.2}$$

9.4 The Neumann problem

We may in a similar way use a Green's function $g(\mathbf{x}, \mathbf{a})$ to solve the corresponding Neumann problem for the Laplace equation, but we must proceed with greater care in this case. It is easy to show that there can exist no function $g(\mathbf{x}, \mathbf{a})$ with the properties

(*a*) $\nabla^2 g(\mathbf{x}, \mathbf{a}) = 0$ in V' (V with the sphere $|\mathbf{x} - \mathbf{a}| = \delta$ removed).
(*b*) $|\mathbf{x} - \mathbf{a}|^{n-2} g(\mathbf{x}, \mathbf{a})$ is bounded and finite near $\mathbf{x} = \mathbf{a}$, and
(*c*) $\partial g/\partial n = 0$ on Σ,

for

$$0 = \int_{V'} \nabla^2 g(\mathbf{x}, \mathbf{a})\, d\tau = \int_{\Sigma} \frac{\partial g}{\partial n} dS - \int_{r=\delta} \frac{\partial g}{\partial r} dS = -\int_{r=\delta} \frac{\partial g}{\partial r} dS$$
$$= (n-2)\Omega_n + O(\delta^2).$$

The difficulty is resolved if we replace condition (*a*) by (*a*′).
(*a*′) $\nabla^2 g(\mathbf{x}, \mathbf{a}) = A$ where A is chosen so that if v is the volume of V

$$Av = (n-2)\Omega_n. \tag{9.4.1}$$

Once again we can find a very simple physical analogue of this paradox. We have mentioned earlier that Neumann problems are typical of hydrodynamics, and so this paradox is equivalent to the statement that there is no solution to the problem posed by placing a simple source in a closed bounded region with fixed walls. In fact it is clear from the identity

$$\int_V \nabla^2 \Phi(\mathbf{x})\, d\tau = \int_\Sigma \frac{\partial \Phi}{\partial n} dS,$$

which follows immediately from the Gauss theorem, that the boundary data of a Neumann problem for the Laplace equation must satisfy

$$\int_\Sigma \frac{\partial \Phi}{\partial n}\, dS = 0.$$

If we can now find the function $g(\mathbf{x}, \mathbf{a})$, and if $\Phi(\mathbf{x})$ satisfies the Poisson equation (6.1.2) we can proceed as we did in Section 9.1. Firstly

$$\int_{V'} \{g(\mathbf{x}, \mathbf{a})\, \nabla^2 \Phi(\mathbf{x}, \mathbf{a}) - \Phi(\mathbf{x})\, \nabla^2 g(\mathbf{x}, \mathbf{a})\}\, d\tau$$
$$= -\Omega_n \int_{V'} \rho(\mathbf{x})\, g(\mathbf{x}, \mathbf{a})\, d\tau - A \int_{V'} \Phi(\mathbf{x})\, d\tau, \tag{9.4.2}$$

and secondly, using the Green identity, the left-hand side is equal to

$$\int_\Sigma g(\mathbf{x}, \mathbf{a}) \frac{\partial \Phi}{\partial n} - \int_{r=\delta} \left\{ g(\mathbf{x}, \mathbf{a}) \frac{\partial \Phi}{\partial r} - \Phi(\mathbf{x}) \frac{\partial g}{\partial r} \right\} dS. \tag{9.4.3}$$

In the limit as $\delta \to 0$

$$(n-2)\Phi(\mathbf{a}) = \int_V \rho(\mathbf{x})\, g(\mathbf{x}, \mathbf{a})\, d\tau - \Omega_n^{-1}\int_\Sigma g(\mathbf{x}, \mathbf{a})\frac{\partial \Phi}{\partial n}\, dS + \frac{n-2}{v}\int_V \Phi(\mathbf{x})\, d\tau. \qquad (9.4.4)$$

The last term on the right-hand side merely appears as an arbitrary constant of integration.

As an example consider the Neumann problem for the Laplace equation inside a circle of radius ρ, then $A = -2/\rho^2$. We then put (using the same notation as in 9.2(b)):

$$g(\mathbf{x}, \mathbf{a}) = -\frac{r^2}{2\rho^2} + \tfrac{1}{2}\log\left[\{(x-\alpha)^2 + (y-\beta)^2\} \times \left\{\left(x - \frac{\rho^2\alpha}{a}\right)^2 + \left(y - \frac{\rho^2\beta}{a}\right)^2\right\}\right] - 2\log\rho. \qquad (9.4.5)$$

The second term is a solution of the Laplace equation, and so condition (a') is satisfied, and it is easily verified that both (b) and (c) are also satisfied. Hence

$$\Phi(\alpha, \beta) = \text{const.} - \frac{a}{2\pi}\int_0^{2\pi}\left(\frac{\partial \Phi}{\partial \rho}\right)_{|x|=\rho} \times \log\left[\{\rho^2 - 2a\rho\cos(\theta - \theta') + a^2\}/\rho^2\right] d\theta' \qquad (9.4.6)$$

which agrees with (6.7.11).

9.5 The expansion of Green's function

Suppose now that $\Phi_\lambda(\mathbf{x})$ is a solution of the Helmholtz equation

$$\nabla^2\Phi_\lambda(\mathbf{x}) + \lambda^2\Phi_\lambda\, \mathbf{x} = 0, \qquad (9.5.1)$$

and if $G(\mathbf{x}, \mathbf{a})$ is the function of Section 9.1, then using the usual identity,

$$\int_{V'}\{\Phi_\lambda(\mathbf{x})\nabla^2 G(\mathbf{x}) - G(\mathbf{x}, \mathbf{a})\nabla^2\Phi_\lambda(\mathbf{x})\}d\tau = \lambda^2\int_{V'} G(\mathbf{x}, \mathbf{a})\,\Phi_\lambda(\mathbf{x})\, d\tau, \qquad (9.5.2)$$

and also

$$\int_\Sigma \Phi_\lambda(\mathbf{x})\frac{\partial G}{\partial n}\, dS - \int_{r=\delta}\left\{\Phi_\lambda(\mathbf{x})\frac{\partial G}{\partial r} - G(\mathbf{x}, \mathbf{a})\frac{\partial \Phi_\lambda}{\partial r}\right\}dS. \qquad (9.5.3)$$

If we can select solutions of (9.5.1) which vanish everywhere on the boundary Σ, then in the limit as $\delta \to 0$

$$\lambda^2\int_V G(\mathbf{x}, \mathbf{a})\,\Phi_\lambda(\mathbf{x})\, d\tau = (n-2)\Omega_n\,\Phi_\lambda(\mathbf{a}). \qquad (9.5.4)$$

This is a homogeneous integral equation for $\Phi_\lambda(\mathbf{x})$ whose kernel is $G(\mathbf{x}, \mathbf{a})$, which is symmetric, as we have shown above. It follows therefore, from the Schmidt theorem [24], that there is at least one eigenvalue for λ for which the equation is soluble. All the eigenvalues must be real. It is certainly true that if V is a rectangular parallelepiped, a sphere or a right circular cylinder, there will be an infinity of eigenvalues, and the corresponding eigenfunctions will form a complete orthogonal set on V. This appears from the cursory treatment that we have given to the Helmholtz equation in Chapter 7. It is beyond the scope of this book to prove more general theorems, but we shall assume for the present that we have an infinite set of real eigenvalues (λ_k) $k = 1, 2, 3$, arranged in ascending order, and the corresponding eigenfunctions $\{\Phi_k(\mathbf{x})\}$ form a complete orthonormal set on V.

The basic properties of such a set are:

$$(a) \int_V \Phi_k(\mathbf{x})\, \Phi_{k'}(\mathbf{x})\, d\tau = 0,\ k \neq k', \tag{9.5.5}$$

$$(b) \int_V \{\Phi_k(\mathbf{x})\}^2\, d\tau = 1 \quad \text{and} \tag{9.5.6}$$

(*c*) if $F(\mathbf{x})$ is an arbitrary function of the vector $\mathbf{x}$, which is absolutely integrable on Ω (that is $\int_V F(\mathbf{x})\, d\tau$ exists and $\int_V |F(\mathbf{x})|\, d\tau < \infty$), then there is an expansion

$$F(\mathbf{x}) = \sum_{k=1}^{\infty} A_k \Phi_k(\mathbf{x}) \tag{9.5.7}$$

where the A_k are the Fourier coefficients

$$A_k = \int_V F(\mathbf{x})\, \Phi_k(\mathbf{x})\, d\tau. \tag{9.5.8}$$

This expansion will converge to $F(\mathbf{x})$ at all interior points of Ω where $F(\mathbf{x})$ is continuous. If $F(\mathbf{x}) = 0$ on some segment of the boundary Σ we can also include this segment in the region of validity of the expansion. The first result (9.5.5) is at any rate made plausible by the usual identity

$$\begin{aligned}(\lambda_k^2 - \lambda_{k'}^2) \int_V \Phi_k(\mathbf{x})\Phi_k\ (\mathbf{x})\, d\tau &= \int_V \{\Phi_k(\mathbf{x})\, \nabla^2\Phi_{k'}(\mathbf{x}) - \Phi_{k'}(\mathbf{x})\, \nabla^2\Phi_k(\mathbf{x})\}\, d\tau \\ &= \int_\Sigma \left\{\Phi_k(\mathbf{x})\frac{\partial\Phi_{k'}}{\partial n} - \Phi_{k'}(\mathbf{x})\frac{\partial\Phi_k}{\partial n}\right\} dS \\ &= 0.\end{aligned}$$

The Green's function $G(\mathbf{x}, \mathbf{a})$ itself will satisfy the restrictive condition of absolute integrability and it vanishes everywhere on Σ. Thus, by its symmetry

$$G(\mathbf{x}, \mathbf{a}) = \sum_{k=1}^{\infty} \gamma_k \Phi_k(\mathbf{x})\, \Phi_k(\mathbf{a}) \tag{9.5.9}$$

and from (9.5.4) and (9.5.7)

$$\gamma_k = \frac{(n-2)\Omega_n}{\lambda_k^{\,2}}. \tag{9.5.10}$$

We may also, of course, seek a Green's function $\Gamma(\mathbf{x}, \mathbf{a}, \lambda)$ for equation (9.5.1). In this case we require conditions (*b*) and (*c*) of 9.1 to be satisfied, but equation (9.5.1) replaces (6.1.1) in the condition (*a*). This function could be used to solve a general Dirichlet type problem for (9.5.1), both for the equation as it stands and for its inhomogeneous equivalent. Since, however, the analysis is almost exactly identical with that of 9.1 above, we shall not elaborate on this.

We also see that

$$\begin{aligned}\lambda^2\int_{V''} & G(\mathbf{x}, \mathbf{a})\,\Gamma(\mathbf{x}, \mathbf{b}, \lambda)\,d\tau \\ &= \int_{V''}\{\Gamma(\mathbf{x}, \mathbf{b}, \lambda)\,\nabla^2 G(\mathbf{x}, \mathbf{a}) - G(\mathbf{x}, \mathbf{a})\,\nabla^2\Gamma(\mathbf{x}, \mathbf{b}, \lambda)\}\,d\tau \\ &= \int_{r=\delta}\left\{G(\mathbf{x}, \mathbf{a})\frac{\partial\Gamma(\mathbf{x}, \mathbf{b}, \lambda)}{\partial r} - \Gamma(\mathbf{x}, \mathbf{b}, \lambda)\frac{\partial G(\mathbf{x}, \mathbf{a})}{\partial r}\right\}dS \\ &\quad + \int_{r'=\delta'}\left\{G(\mathbf{x}, \mathbf{a})\frac{\partial\Gamma(\mathbf{x}, \mathbf{b}, \lambda)}{\partial r'} - \Gamma(\mathbf{x}, \mathbf{b}, \lambda)\frac{\partial G(\mathbf{x}, \mathbf{a})}{\partial r'}\right\}dS.\end{aligned}$$

Hence in the limit

$$\lambda^2\int_V G(\mathbf{x}, \mathbf{a})\,\Gamma(\mathbf{x}, \mathbf{b}, \lambda)\,d\tau = (n-2)\Omega_n\{\Gamma(\mathbf{a}, \mathbf{b}, \lambda) - G(\mathbf{b}, \mathbf{a})\}. \tag{9.5.11}$$

The symmetry of $\Gamma(\mathbf{x}, \mathbf{b}, \lambda)$ follows obviously from this relation. We notice also that $\Gamma(\mathbf{x}, \mathbf{b}, \lambda)$ can be regarded as the solution of an inhomogeneous integral equation of which $G(\mathbf{x}, \mathbf{a})$ is kernel, and vice versa. In the terminology of the theory of integral equations each function is the resolvent of the other. Finally, if λ is not an eigenvalue of (9.5.4), we may write:

$$\Gamma(\mathbf{x}, \mathbf{b}, \lambda) = \sum_{k=0}^{\infty}\Gamma_k\Phi_k(\mathbf{x})\,\Phi_k(\mathbf{b}), \tag{9.5.12}$$

where the coefficients Γ_k are functions of λ. Substitution into (9.5.11) gives

$$\lambda^2\sum_{k=1}^{\infty}\Gamma_k\gamma_k\Phi_k(\mathbf{a})\,\Phi_k(\mathbf{b}) = (n-2)\Omega_n\sum_{k=1}^{\infty}(\Gamma_k - \gamma_k)\,\Phi_k(\mathbf{a})\,\Phi_k(\mathbf{b}).$$

If now we equate coefficients on both sides of the equation:

$$(\lambda_k^{\,2} - \lambda^2)\Gamma_k = \lambda_k^{\,2}\gamma_k = (n-2)\Omega_n. \tag{9.5.13}$$

9.6 The general linear elliptic equation of second order

We can generalize the idea of a Green's function to other elliptic operators, if we make use of the notions of Section 5.5. Let us suppose that we seek a solution of the elliptic equation:

$$L(\Phi) = A_{ij}(\mathbf{x})\,\Phi_{ij} + B_i(\mathbf{x})\,\Phi_i + C(\mathbf{x})\,\Phi = f(\mathbf{x}) \qquad (9.6.1)$$

in the interior of the closed bounded region V, with boundary Σ, given the value of $\Phi(\mathbf{x})$ on the boundary Σ.

Since we need to use the equivalent of the Green's identity (5.5.1), the required Green's function must be a solution of the adjoint equation

$$M\{\Psi\} = \frac{\partial^2}{\partial x_i\,\partial x_j}\{A_{ij}(\mathbf{x})\,\Psi(\mathbf{x})\} - \frac{\partial}{\partial x_i}\{B_i(\mathbf{x})\,\Psi(x)\} + C(\mathbf{x})\,\Psi(x) = 0. \qquad (9.6.2)$$

The function $G(\mathbf{x}, \mathbf{a})$ will in fact have the three properties corresponding to those of Section 9.1:

(*a*) $G(\mathbf{x}, \mathbf{a})$ is a solution of (9.6.2) at every interior point of V except $\mathbf{a}$.
(*b*) Near $\mathbf{x} = \mathbf{a}$, $n\,|\,\mathbf{x} - \mathbf{a}\,|^{n-2}\,G(\mathbf{x}, \mathbf{a})/A_{ii}(\mathbf{a})$ approaches unity,
(*c*) $G(\mathbf{x}, \mathbf{a})$ vanishes on the boundary Σ.

Now we can proceed as before, using the identity (5.5.1) to replace that of Green, and so

$$\begin{aligned}\int_{V'} G(\mathbf{x}, \mathbf{a})\,f(\mathbf{x})\,d\tau &= \int_{V'} [G(\mathbf{x}, \mathbf{a})L(\Phi) - \Phi(\mathbf{x})\,M\{G(\mathbf{x}, \mathbf{a})\}]\,d\tau \\ &= -\int_{\Sigma} A_{ij}(\mathbf{x})\,n_j\frac{\partial G(\mathbf{x}, \mathbf{a})}{\partial x_i}\Phi(\mathbf{x})\,dS \\ &\quad - \int_{r=\delta}\Bigg[A_{ij}(\mathbf{x})\left\{n_i G(\mathbf{x}, \mathbf{a})\frac{\partial \Phi}{\partial \mathbf{x}_j} - n_j\Phi(\mathbf{x})\frac{\partial G(\mathbf{x}, \mathbf{a})}{\partial x_i}\right\} \\ &\quad + \left\{n_i\,B_i\,(\mathbf{x}) - n_j\frac{\partial A_{ij}(\mathbf{x})}{\partial x_i}\right\}G(\mathbf{a}, \mathbf{x})\,\Phi(\mathbf{x})\Bigg]\,dS.\end{aligned}$$

In the limit, the only term in the last integral which will contribute is that containing derivatives of $G(\mathbf{x}, \mathbf{a})$. Furthermore

$$\frac{\partial G(\mathbf{x}, \mathbf{a})}{\partial x_i} = \frac{\partial G(\mathbf{x}, \mathbf{a})}{\partial r}n_i$$

and

$$\lim_{\delta\to 0} \delta^{1-n}\int_{r=\delta} A_{ij}(\mathbf{x})\,n_i\,n_j\,dS = \frac{(n-2)}{n}\Omega_n A_{ii}(\mathbf{a}).$$

[NOTE: Since we have an elliptic operator we may assume $A_{ij}\,n_i\,n_j > 0$ and so therefore is A_{ii}].

Finally therefore:

$$\int_V G(\mathbf{x}, \mathbf{a}) f(\mathbf{x})\, d\tau + \int_\Sigma \frac{\partial G(\mathbf{x}, \mathbf{a})}{\partial n} \Phi(\mathbf{x})\, A_{ij}(\mathbf{x})\, n_i n_j\, dS + (n-2)\Omega_n \Phi(\mathbf{a}) = 0. \qquad (9.6.3)$$

In the general case there will not be any question of symmetry in the function $G(\mathbf{x}, \mathbf{a})$, and without this symmetry we could not use the theorems on integral equations, which are needed to generalize the analysis of the previous section. If, however, the equation is self-adjoint, that is if it can be written:

$$L(\Phi) \equiv \frac{\partial}{\partial x_j}\left\{A_{ij}(\mathbf{x})\frac{\partial \Phi}{\partial x_i}\right\} + C(\mathbf{x})\, \Phi(\mathbf{x}) = f(x), \qquad (9.6.4)$$

then $G(\mathbf{x}, \mathbf{a})$ will have the necessary properties of symmetry. We may then consider, with (9.6.4), the second equation:

$$L\{\Psi(\mathbf{x})\} + \lambda^2 d(\mathbf{x})\, \Psi(\mathbf{x}) = 0 \qquad (9.6.5)$$

wherein we assume that $d(\mathbf{x}) > 0$ for all $\mathbf{x} \in V$.

Then if $\Psi_\lambda(\mathbf{x})$ is a solution vanishing everywhere on Σ,

$$\begin{aligned}\lambda^2 \int_V d(\mathbf{x}) G(\mathbf{x}, \mathbf{a})\, \Psi_\lambda(\mathbf{x})\, d\tau &= \lim_{\delta \to 0} \int_{V'} [\Psi_\lambda(\mathbf{x})\, L\{G(\mathbf{x}, \mathbf{a})\} \\ &\qquad - G(\mathbf{x}, \mathbf{a})\, L\{\Psi_\lambda(\mathbf{x})\}]\, d\tau \\ &= (n-2)\Omega_n \Psi_\lambda(\mathbf{a}). \end{aligned} \qquad (9.6.6)$$

Hence, as in the previous section, we may produce a set of functions $\{\Psi_k(\mathbf{x})\}$, corresponding to the eigenvalues λ_k and these functions can be normalized so that

$$\int_V d(\mathbf{x})\, \Psi_k(\mathbf{x})\, \Psi_{k'}(\mathbf{x})\, d\tau = 0, \quad k \neq k' \qquad (9.6.7)$$

and

$$\int_V d(\mathbf{x})\, \{\Psi_k(\mathbf{x})\}^2\, d\tau = 1. \qquad (9.6.8)$$

In particular, corresponding to (9.5.9) we shall find that

$$G(\mathbf{x}, \mathbf{a}) = (n-2)\Omega_n \sum_{k=1}^{\infty} \lambda_k^{-2}\, \Psi_k(\mathbf{x})\, \Psi_k(\mathbf{a}). \qquad (9.6.9)$$

All this is clearly in direct analogy with the ordinary Sturm-Liouville theory.

Exercises

9.1. Show that the transformation

$$z = x + iy = \sqrt{\zeta} = (\xi + i\eta)^{1/2}$$

(where the principal value of the root is understood) maps the half-plane $\mathscr{R}(\zeta) \geqslant 0$ on the right-angled segment $|\arg z| \leqslant \pi/4$. Hence or otherwise find the two-dimensional Green's function for the Laplace equation, when the boundary of the region is the two half-lines $x \pm y = 0$, $x > 0$ and the region is $|y| < |x|$.

9.2. Show that the transformation

$$x + iy = z = \sinh^{-1} \zeta = \sinh^{-1} \{\xi + i\eta\}$$

maps the half-plane $\mathscr{R}(\zeta) = 0$ on to the semi-infinite strip $x \geqslant 0$, $-\pi/2 \leqslant y \leqslant \pi/2$. Hence or otherwise find the two-dimensional Green's function of the Laplace equation for this bounded region.

9.3. Show that a 'point charge' in front of the surface consisting of the plane $z = 0$ with the hemispherical boss $x^2 + y^2 + z^2 = \rho^2$, $z > 0$, when the surface is earthed, is equivalent to an image system of four charges. Hence or otherwise find the Green's function for this surface.

9.4. Show that a solution of

$$L(\Phi) \equiv \frac{\partial^2 \Phi}{\partial x^2} + \frac{\partial^2 \Phi}{\partial y^2} + \frac{\partial^2 \Phi}{\partial z^2} - k^2 \Phi = 0$$

everywhere excluding a neighbourhood of (x', y', z') and which $\rightarrow 0$ as $r \rightarrow \infty$ is e^{-kr}/r, where $r^2 = (x - x')^2 + (y - y')^2 + (z - z')^2$.

Deduce that if V is a simply-connected, bounded region of three-dimensional space, bounded by a regular surface Σ, and if $f(x, y, z)$ is integrable over V then

$$F(x', y', z') = \int_V f(x, y, z)\{e^{-kr}/r\}\, dx\, dy\, dz$$

is a solution of

$$\begin{aligned} L(\Phi) &= f(x, y, z) \text{ if } (x, y, z) \text{ is within } V \\ &= 0 \qquad\quad\;\; \text{if } (x, y, z) \text{ is outside } V, \end{aligned}$$

is continuous and with continuous normal derivative across Σ, and vanishing at ∞.

9.5. Find a Green's function for the equation $L(\Phi) = 0$ of Exercise 9.4, for the region $z > 0$ bounded by the plane $z = 0$.

ANSWERS

9.1. $G(\mathbf{x}, \mathbf{a}) = \frac{1}{2} \log[\{(x^2 - y^2 - \alpha^2 + \beta^2)^2 + 4(xy - \alpha\beta)^2\}/\{(x^2 - y^2 + \alpha^2 - \beta^2)^2 + 4(xy - \alpha\beta)^2\}]$.

9.2. $G(\mathbf{x}, \mathbf{a}) = \frac{1}{2}[\log\{(\sinh x \cos y - \sinh \alpha \cos \beta)^2 + (\cosh x \sin y - \cosh \alpha \sin \beta)^2\} - \log\{(\sinh x \cos y + \sinh \alpha \cos \beta)^2 + (\cosh x \sin y - \cosh \alpha \sin \beta)^2\}]$.

9.3. $G(\mathbf{x}, \mathbf{a}) = \left[\frac{1}{r} - \frac{1}{r_1} - \lambda\left\{\frac{1}{r_2} - \frac{1}{r_3}\right\}\right]$

where $a\lambda = \rho$, $a^2 = |\mathbf{a}|^2 = \alpha^2 + \beta^2 + \gamma^2$, $r^2 = |\mathbf{x} - \mathbf{a}|^2$,

$r_1{}^2 = (x + \alpha)^2 + (y - \beta)^2 + (z - \gamma)^2$,

$r_2{}^2 = (x - \lambda^2\alpha)^2 + (y - \lambda^2\beta)^2 + (z - \lambda^2\gamma)^2$

and $r_3{}^2 = (x + \lambda^2\alpha)^2 + (y - \lambda^2\beta)^2 + (z - \lambda^2\gamma)^2$.

(9.5. $G(\mathbf{x}, \mathbf{a}) = \{\exp(-kr)\}/r - \{\exp(-kr_1)\}/r_1$ where r_1 is defined above.

CHAPTER 10

The Riemann Theory of the Hyberbolic Equations

10.1 The equation in two independent variables

Let us consider the canonical form for the linear second-order hyperbolic equation in two independent variables:

$$L(\Phi) = A(x, y)\frac{\partial^2\Phi}{\partial x\,\partial y} + B_1(x, y)\frac{\partial\Phi}{\partial x} + B_2(x, y)\frac{\partial\Phi}{\partial y} + C(x, y)\,\Phi(x, y) = f(x, y). \qquad (10.1.1)$$

From the Cauchy-Kowalewski existence theorem we see that if the coefficients are sufficiently well behaved, then, given as boundary data values of $\Phi(x, y)$ and $\partial/\partial n\{\Phi(x, y)\}$ on an arc of a smooth curve C which

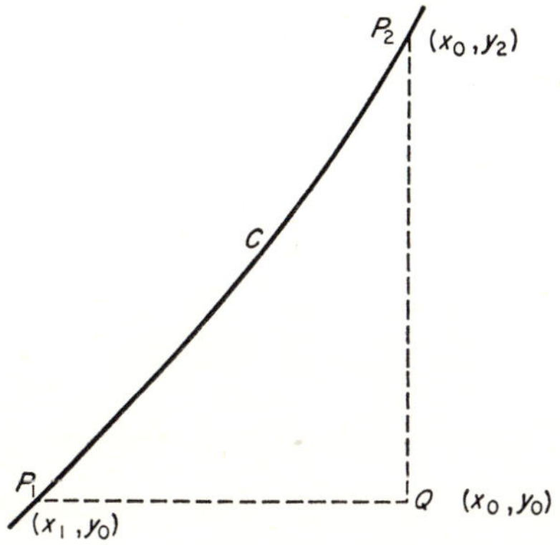

Fig. 10.1

is nowhere characteristic, we can determine the solution of (10.1.1) in a triangular region adjacent to the curve. If we refer to Fig. 10.1, we can say that if the arc extends from the point $P_1(x_1, y_0)$ to $P_2(x_0, y_2)$, then the solution is determined in the region between C and the lines P_1Q, QP_2, where Q has coordinates (x_0, y_0). This is because the characteristics of (10.1.1) are lines $x =$ const. and $y =$ const., and so Q is the point where a characteristic of one family through P_1 meets a member of the other family through P_2. Let us denote the enclosed area by Δ.

The adjoint equation to (10.1.1) is

$$M(\Psi) = \frac{\partial^2}{\partial x\,\partial y}\{A(x, y)\,\Psi(x, y)\} - \frac{\partial}{\partial x}\{B_1(x, y)\,\Psi(x, y)\}$$
$$- \frac{\partial}{\partial y}\{B_2(x, y)\,\Psi(x, y)\} + C(x, y)\,\Psi(x, y)\} = 0. \qquad (10.1.2)$$

For the moment, let $\Psi(x, y)$ represent any solution of (10.1.2) in Δ with continuous first derivatives. Then

$$2\int_\Delta \Psi(x, y)\, f(x, y)\, d\tau$$
$$= 2\int_\Delta [\Psi(x, y)\, L\{\Phi(x, y)\} - \Phi(x, y)\, M\{\Psi(x, y)\}]\, d\tau$$
$$= \int_{y_0}^{y_2}\Bigg[A(x, y)\left\{\Psi(x, y)\frac{\partial \Phi}{\partial y} - \Phi(x, y)\frac{\partial \Psi}{\partial y}\right\}$$
$$+ \left\{2B_1(x, y) - \frac{\partial A}{\partial y}\right\}\Phi(x, y)\,\Psi(x, y)\Bigg]_{x=x_0} dy$$
$$- \int_{x_1}^{x_0}\Bigg[A(x, y)\left\{\Psi(x, y)\frac{\partial \Phi}{\partial x} - \Phi(x, y)\frac{\partial \Psi}{\partial x}\right\}$$
$$+ \left\{2B_2(x, y) - \frac{\partial A}{\partial x}\right\}\Phi(x, y)\,\Psi(x, y)\Bigg]_{y=y_0} dx$$
$$+ \int_C\Bigg[A(x, y)\left\{\Psi(x, y)\frac{\partial \Phi}{\partial \nu} - \Phi(x, y)\frac{\partial \Psi}{\partial \nu}\right\}$$
$$+ \left\{2B_\nu(x, y) - \frac{\partial A}{\partial \nu}\right\}\Phi(x, y)\,\Psi(x, y)\Bigg]\, ds. \qquad (10.1.3)$$

The last integral is a curvilinear integral taken from P_1 to P_2 along C. Also

$$\frac{\partial}{\partial \nu} = \frac{\partial}{\partial x}\frac{dx}{ds} - \frac{\partial}{\partial y}\frac{dy}{ds}$$

and

$$B_\nu(x, y) = B_2(x, y)\frac{dx}{ds} - B_1(x, y)\,\frac{dy}{ds}.$$

In both the first and second term on the right-hand side of (10.1.3) we can integrate by parts so that

$$\int_{y_0}^{y_2}\left\{A(x, y)\,\Psi(x, y)\frac{\partial \Phi}{\partial y}\right\}_{x=x_0} dy$$
$$= (A\Phi\Psi)_2 - (A\Phi\Psi)_Q - \int_{y_0}^{y_2}\left\{\Phi(x, y)\frac{\partial}{\partial y}(A\Psi)\right\}_{x=x_0} dy$$

and

$$\int_{x_1}^{x_0}\left\{A\Psi\frac{\partial\Phi}{\partial x}\right\}_{y=y_0} dx = (A\Phi\Psi)_Q - (A\Phi\Psi)_1 - \int_{x_1}^{x_0}\left\{\Phi(x, y)\frac{\partial}{\partial x}(A\Psi)\right\}_{y=y_0} dx$$

where suffices 2 and 1 denote values at P_2 and P_1.

With these modifications, the integrand of the integral taken along the line $x = x_0$ is just

$$2\Phi(x, y)\left\{B_1\Psi - \frac{\partial}{\partial y}(A\Psi)\right\}$$

and similarly the integral along $y = y_0$ has the integrand

$$2\Phi(x, y)\left\{\frac{\partial}{\partial x}(A\Psi) - B_2\Psi\right\}.$$

We now therefore particularize our choice of the function $\Psi(x, y)$. The resulting function which we shall call the Riemann function $R(x, y, x_0, y_0)$ has the properties:

(*a*) at every interior point of Δ, $M\{R(x, y, x_0, y_0)\} = 0$

(*b*) on the line $x = x_0$

$$A(x_0, y)\, R(x_0, y, x_0, y_0) \propto \exp\left[\int^{y}\{B_1(x_0, y)/A(x_0, y)\}\, dy\right], \text{ and}$$

(*c*) on the line $y = y_0$

$$A(x, y_0)\, R(x, y_0, x_0, y_0) \propto \exp\left[\int^{x}\{B_2(x, y_0)/A(x, y_0)\}\, dx\right].$$

Then

$$2\int_{\Delta} R(x, y, x_0, y_0) f(x, y)\, d\tau = \{A\Phi R\}_{P_2} + \{A\Phi R\}_{P_1} - 2\{A\Phi R\}_Q$$
$$+ \int_C \left[A\left\{R\frac{\partial\Phi}{\partial\nu} - \Phi\frac{\partial R}{\partial\nu}\right\} + \left\{2B_\nu - \frac{\partial A}{\partial\nu}\right\} R\Phi\right] ds. \qquad (10.1.4)$$

This expresses the solution of the equation (10.1.1) at Q as a sum of four terms, of which the first is a known integral over Δ, the next two involve only values of Φ at P_1 and P_2, and the last a line integral along C. Although in this last integral the partial derivatives $\partial/\partial\nu$ occur, these can be found since we know both the tangential and normal derivatives of Φ along C.

The Riemann function depends both on the coordinates of the point Q, (x_0, y_0), and on the running coordinates x, y. As in the case of the Green's function, it will only be symmetric in the two sets of coordinates if equation (10.1.1) is self-adjoint. This condition is, however, realized in a number of equations of interest.

10.2 Some illustrative examples

We shall consider three examples:

(*a*) The wave equation

$$\frac{\partial^2 \Phi}{\partial x\, \partial y} = f(x, y). \tag{10.2.1}$$

This is self-adjoint, and the conditions (*b*) and (*c*) merely reduce to $R(x, y, x_0, y_0)$ is constant on the lines $x = x_0$, $y = y_0$.

It is therefore sufficient to put

$$R(x, y, x_0, y_0) = 1. \tag{10.2.2}$$

This will allow (10.1.4) to be written as

$$\Phi(x_0, y_0) = \tfrac{1}{2}\Phi_{P_1} + \tfrac{1}{2}\Phi_{P_2} + \tfrac{1}{2}\int_C \frac{\partial \Phi}{\partial \nu} ds - \int_\Delta f(x, y)\, d\tau. \tag{10.2.3}$$

(*b*) The equation

$$4\frac{\partial^2 \Phi}{\partial x\, \partial y} + \Phi = f(x, y). \tag{10.2.4}$$

This again is a self-adjoint equation, and as in (*a*) the boundary conditions for the Riemann function are that it is constant on $x = x_0$ and $y = y_0$. We can, however, no longer satisfy (10.2.4) with a constant value.

Let

$$\eta = \{(x - x_0)(y - y_0)\}^{1/2},$$

and seek a solution $R(x, y, x_0, y_0) = R(\eta)$, since this will have the desired symmetry and will reduce to $R(0) = \text{const.}$ on the boundary.

Now

$$\frac{\partial^2 R}{\partial x\, \partial y} = \frac{d^2 R}{d\eta^2}\frac{\partial \eta}{\partial x}\frac{\partial \eta}{\partial y} + \frac{dR}{d\eta}\frac{\partial^2 \eta}{\partial x\, \partial y} = \frac{1}{4}\left\{\frac{d^2 R}{d\eta^2} + \frac{1}{\eta}\frac{dR}{d\eta}\right\},$$

and so

$$\eta\frac{d^2 R}{d\eta^2} + \frac{dR}{d\eta} + \eta R = 0. \tag{10.2.5}$$

Hence we may take

$$R(x, y, x_0, y_0) = J_0[\{(x - x_0)(y - y_0)\}^{1/2}]. \tag{10.2.6}$$

Then

$$\Phi(x_0 y_0) = \tfrac{1}{2}\Phi_{P_1} + \tfrac{1}{2}\Phi_{P_2} + \tfrac{1}{2}\int_C \left\{J_0(\eta)\frac{\partial \Phi}{\partial \nu} - \Phi\frac{\partial J_0}{\partial \nu}\right\} ds - \int_\Delta J_0(\eta) f(x, y)\, d\tau. \tag{10.2.7}$$

(*c*) Finally we consider an equation of importance in modern gas dynamics, the so-called Darboux equation

$$(x + y)\frac{\partial^2 \Phi}{\partial x\, \partial y} + k\left\{\frac{\partial \Phi}{\partial x} + \frac{\partial \Phi}{\partial y}\right\} = f(x, y), \quad (k > 1). \tag{10.2.8}$$

This is not self-adjoint, although it could be made so by multiplying through by $(x + y)^{k-1}$. The adjoint equation is

$$(x + y)\frac{\partial^2 \Psi}{\partial x\, \partial y} + (1 - k)\left\{\frac{\partial \Psi}{\partial x} + \frac{\partial \Psi}{\partial y}\right\} = 0. \tag{10.2.9}$$

Thus $R(x, y, x_0, y_0)$ is a solution of (10.2.9) which is proportional to $(x_0 + y)^{k-1}$ on $x = x_0$ and to $(x + y_0)^{k-1}$ on $y = y_0$. We look for a solution

$$R(x, y, x_0, y_0) = \left(\frac{x + y}{x_0 + y_0}\right)^{k-1} \rho(\eta) \tag{10.2.10}$$

where

$$\eta = \frac{(x - x_0)(y - y_0)}{(x + y)(x_0 + y_0)}.$$

Thus

$$\frac{\partial R}{\partial x} = \left(\frac{x + y}{x_0 + y_0}\right)^{k-1}\left\{\frac{d\rho}{d\eta}\frac{\partial \eta}{\partial x} + \frac{(k - 1)\rho}{x + y}\right\},$$

$$\frac{\partial R}{\partial y} = \left(\frac{x + y}{x_0 + y_0}\right)^{k-1}\left\{\frac{d\rho}{d\eta}\frac{\partial \eta}{\partial y} + \frac{(k - 1)\rho}{x + y}\right\},$$

and

$$\frac{\partial^2 R}{\partial x\, \partial y} = \left(\frac{x + y}{x_0 + y_0}\right)^{k-1}\left[\frac{d^2\rho}{d\eta^2}\frac{\partial \eta}{\partial x}\frac{\partial \eta}{\partial y} + \frac{d\rho}{d\eta}\left\{\frac{\partial^2 \eta}{\partial x\, \partial y} + \frac{(k - 1)}{x + y}\left(\frac{\partial \eta}{\partial x} + \frac{\partial \eta}{\partial y}\right)\right\} + \frac{(k - 1)(k - 2)\rho}{(x + y)^2}\right],$$

so that

$$\frac{d^2\rho}{d\eta^2}\frac{\partial \eta}{\partial x}\frac{\partial \eta}{\partial y} + \frac{d\rho}{d\eta}\frac{\partial^2 \eta}{\partial x\, \partial y} - \frac{k(k - 1)}{(x + y)^2}\rho = 0. \tag{10.2.11}$$

On the other hand

$$\frac{1}{\eta}\frac{\partial \eta}{\partial x} = \frac{1}{x - x_0} - \frac{1}{x + y} = \frac{y + x_0}{(x - x_0)(x + y)},$$

$$\frac{\partial \eta}{\partial x}\frac{\partial \eta}{\partial y} = \frac{(y + x_0)(x + y_0)}{(x - x_0)(y - y_0)}\frac{\eta^2}{(x + y)^2}$$

$$= \frac{(y + x_0)(x + y_0)}{(x_0 + y_0)(x + y)}\frac{\eta}{(x + y)^2} = \frac{\eta(1 + \eta)}{(x + y)^2}$$

while

$$\frac{\partial^2 \eta}{\partial x\, \partial y} = \frac{y + x_0}{(x - x_0)(x + y)}\frac{\partial \eta}{\partial y} + \frac{\eta}{(x - x_0)(x + y)} - \frac{(y + x_0)\eta}{(x - x_0)(x + y)^2}$$

$$= \frac{\eta}{(x + y)^2}\left\{\frac{(y + x_0)(x + y_0)}{(x - x_0)(y - y_0)} + 1\right\} = \frac{2\eta + 1}{(x + y)^2}.$$

When the substitutions are made into it, (10.2.11) reduces to

$$\eta(1+\eta)\frac{d^2\rho}{d\eta^2}+(1+2\eta)\frac{d\rho}{d\eta}-k(k-1)\rho(\eta)=0, \qquad (10.2.12)$$

which is a hypergeometric equation. In particular, the solution which reduces to unity when $\eta = 0$ is:

$$\rho(\eta) = F(k, 1-k; 1; -\eta) \qquad (10.2.13)$$

or when k is an integer and use is made of Murphy's formula expressing the Legendre functions of the first kind as hypergeometric functions (see [24])

$$\rho(\eta) = P_{k-1}(1+2\eta). \qquad (10.2.14)$$

Finally then

$$R(x, y, x_0, y_0) = \left(\frac{x+y}{x_0+y_0}\right)^{k-1} F\left\{k, 1-k; 1; -\frac{(x-x_0)(y-y_0)}{(x+y)(x_0+y_0)}\right\} \qquad (10.2.15)$$

and this can be used in (10.1.4) to find the solution of (10.2.8).

10.3 The equation in *n* independent variables

We would like to generalize these ideas, to apply them to linear partial differential equations of normal hyperbolic type in several variables. Although we shall treat the equation in n variables, there are no geometric ideas in the analysis which are not exemplified in full

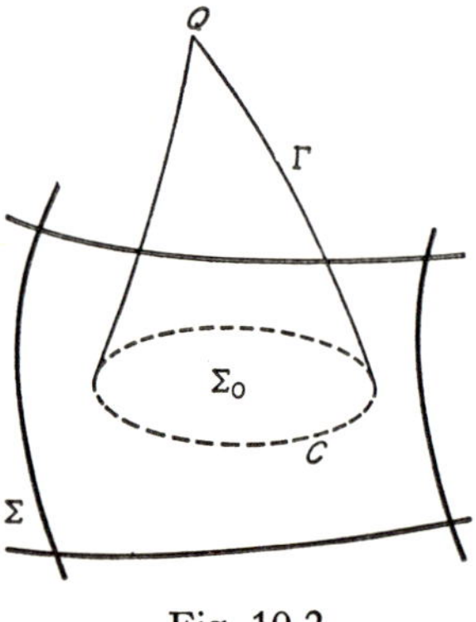

Fig. 10.2

when there are three variables, and so the diagrams we shall draw are for a three-dimensional space (see Fig. 10.2).

Since our partial differential equation

$$L(\Phi) \equiv A_{ij}(\mathbf{x})\,\Phi_{ij} + B_i(\mathbf{x})\,\Phi_i + C(\mathbf{x})\,\Phi(\mathbf{x}) = f(\mathbf{x}) \qquad (10.3.1)$$

is of normal hyperbolic type, then through each point Q of space will pass a non-degenerate characteristic conoid Γ. We assume that this makes an intercept on an $(n-1)$-dimensional hypersurface Σ, on which are known values of $\Phi(\mathbf{x})$ and $\partial\Phi/\partial n$, its derivative in the direction normal to the surface. The three-dimensional analogue suggests that we call the intersection of Γ and Σ the twisted hypercurve C, and the part of Σ contained within C we shall denote by Σ_0. Then the solution of (10.3.1) at the point Q is entirely determined by the values of the boundary data within Σ_0 and on C.

As in Section 10.1, let us assume temporarily that $\Psi(\mathbf{x})$ is any solution of the adjoint equation:

$$M(\Psi) \equiv \frac{\partial^2}{\partial x_i\,\partial x_j}\{A_{ij}(\mathbf{x})\,\Psi(\mathbf{x})\} - \frac{\partial}{\partial x_i}\{B_i(\mathbf{x})\,\Psi(\mathbf{x})\} + C(\mathbf{x})\,\Psi(\mathbf{x}) = 0. \tag{10.3.2}$$

We integrate over that part of the space V contained between Γ and Σ_0. Thus

$$\begin{aligned}\int_V \Psi(\mathbf{x})\,f(\mathbf{x})\,d\tau &= \int_V \{\Psi L(\Phi) - \Phi M(\Psi)\}\,d\tau \\ &= \int_\Gamma \left\{\Lambda\left(\Psi\frac{\partial\Phi}{\partial\nu} - \Phi\frac{\partial\Psi}{\partial\nu}\right) + n_i\left(B_i - \frac{\partial A_{ij}}{\partial x_i}\right)\Psi\Phi\right\} dS \\ &\quad + \int_{\Sigma_0} \left\{\Lambda\left(\Psi\frac{\partial\Phi}{\partial\nu} - \Phi\frac{\partial\Psi}{\partial\nu}\right) + n_i\left(B_i - \frac{\partial A_{ij}}{\partial x_j}\right)\Psi\Phi\right\} dS\end{aligned} \tag{10.3.3}$$

making use of (5.5.2). As in that section, $\partial/\partial\nu$ denotes differentiation, in the direction of the conormal. Now it was shown there that the conormal direction was always tangent to a bicharacteristic direction at the point, and so on Γ the differentiation is along a direction tangent to the surface.

As in the two-variable problem we concentrate our attention on the single term:

$$\int_\Gamma \Lambda\Psi\frac{\partial\Phi}{\partial\nu}\,dS.$$

The surface Γ is an $(n-1)$-dimensional manifold, containing the direction $\boldsymbol{\nu}$. We can thus take a system of curvilinear coordinates $\eta_0, \eta_1, \ldots, \eta_{n-2}$ on the surface, so that the direction η_0 increasing is that of the vector $\boldsymbol{\nu}$. By introducing suitable scaling we can put

$$d\eta_0 = d\nu$$

and assume that $\eta_0 = 0$ at Q. If then D denotes the requisite Jacobian, on Γ,

$$dS = D\,d\nu\,d\eta_1\,d\eta_2\ldots d\eta_{n-2}.$$

Hence

$$\int_\Gamma \Lambda\Psi\frac{\partial\Phi}{\partial\nu}\,dS = \int d\eta_1\,d\eta_2\ldots d\eta_{n-2}\int \Lambda\Psi\frac{\partial\Phi}{\partial\nu}D\,d\nu,$$

$$= \Delta\int D\Lambda\Psi\Phi d\eta_1,\,d\eta_2\ldots d\eta_{n-2} - \int D^{-1}\Phi\frac{\partial}{\partial\nu}(\Lambda D\Psi)\,dS. \tag{10.3.4}$$

The symbol Δ applied to the first integral implies that we have the difference of two terms, taken at the extreme values of η_0. At the upper extreme this will be an integration over the $(n-2)$-dimensional manifold wherein the conoid Γ meets Σ, that is our hypercurve C. At the lower extreme, which is $\eta_0 = 0$, we have an integration over the $(n-2)$-dimensional manifold which is the section by $\eta_0 = 0$ of Γ, unless D vanishes. In two dimensions, this will give actual functional values at Q, as we saw in the preceding sections. In spaces of a greater number of dimensions, this will give, not the value of $\Phi(\mathbf{x})$ at Q, but at best an integral related to it.

We are still left with an integral over the sides of the conoid Γ, namely

$$-\int_\Gamma \Phi\left[2\Lambda\frac{\partial\Psi}{\partial\nu} + \left\{\Lambda D^{-1}\frac{\partial D}{\partial\nu} + \frac{\partial\Lambda}{\partial\nu} - B_i n_i - \frac{\partial A_{ij}}{\partial x_j}n_i\right\}\Psi\right]dS, \tag{10.3.5}$$

and we want this to vanish identically. Our Riemann function would therefore be chosen with the properties that:

(*a*) It is a solution of (10.3.2).

(*b*) On the characteristic conoid

$$2\Lambda D\frac{\partial\Psi}{\partial\nu} = \left\{\left(B_i n_i + \frac{\partial A_{ij}}{\partial x_j}n_i - \frac{\partial\Lambda}{\partial\nu}\right)D - \Lambda\frac{\partial D}{\partial\nu}\right\}\Psi.$$

Formula (10.3.3) would not give directly the value of $\Phi(\mathbf{x})$ at Q, but a related integral as we have seen above.

10.4 Volterra's solution of the wave equation in three variables

We shall apply these ideas to the wave equation, as was first done by Volterra.

$$\frac{\partial^2\Phi}{\partial x^2} + \frac{\partial^2\Phi}{\partial y^2} - \frac{\partial^2\Phi}{\partial z^2} = f(x, y, z). \tag{10.4.1}$$

The characteristic conoid Γ through the point (x_0, y_0, z_0) coincides with the characteristic cone and has the equation

$$(x - x_0)^2 + (y - y_0)^2 - (z - z_0)^2 = 0. \tag{10.4.2}$$

At a general point (x, y, z) on this cone the normal vector $\mathbf{n}$ has components

$$\mathbf{n} = \left\{\frac{x - x_0}{r}, \frac{y - y_0}{r}, \frac{-(z - z_0)}{r}\right\} \tag{10.4.3}$$

where $r^2 = (x - x_0)^2 + (y - y_0)^2 + (z - z_0)^2$, and the vector $\boldsymbol{\nu}$ has components

$$\boldsymbol{\nu} = \left(\frac{x - x_0}{r}, \frac{y - y_0}{r}, \frac{z - z_0}{r}\right) = \boldsymbol{\Lambda}.$$

Hence

$$\frac{\partial \Phi}{\partial \nu} = \frac{x - x_0}{r}\frac{\partial \Phi}{\partial x} + \frac{y - y_0}{r}\frac{\partial \Phi}{\partial y} + \frac{z - z_0}{r}\frac{\partial \Phi}{\partial z}.$$

On Γ we can put

$$x = x_0 + \sigma \cos \theta,\ y = y_0 + \sigma \sin \theta,\ z = z_0 - \sigma,$$

and in terms of these coordinates

$$dS = \sqrt{2}\sigma\, d\sigma\, d\theta,$$

and

$$\sqrt{2}\frac{\partial \Phi}{\partial \nu} = \frac{\partial \Phi}{\partial \sigma}.$$

We shall suppose that the intersection of the cone with Σ is given by

$$\sigma = \sigma_0(\theta).$$

Then equation (10.3.3) reads

$$\int_V \Psi(\mathbf{x}) f(\mathbf{x})\, d\tau = \int_\Gamma \left\{\Psi\frac{\partial \Phi}{\partial \nu} - \Phi\frac{\partial \Psi}{\partial \nu}\right\} dS + \int_{\Sigma_0}\left(\Psi\frac{\partial \Phi}{\partial \nu} - \Phi\frac{\partial \Psi}{\partial \nu}\right) dS. \tag{10.4.4}$$

We may write the first term on the right-hand side as:

$$\int_0^{2\pi} d\theta \int_0^{\sigma_0(\theta)}\left(\Psi\frac{\partial \Phi}{\partial \sigma} - \Phi\frac{\partial \Psi}{\partial \sigma}\right)\sigma\, d\sigma = \int_0^{2\pi} \{(\sigma\Psi\Phi)_{\sigma=\sigma_0(\theta)} - (\sigma\Psi\Phi)_{\sigma=0}\}\, d\theta$$
$$- \int_0^{2\pi} d\theta \int_0^{\sigma_0(\theta)} \Phi\left(2\sigma\frac{\partial \Psi}{\partial \sigma} + \Psi\right) d\sigma. \tag{10.4.5}$$

We want, of course, the contribution from the last integral to vanish identically. This suggests that we want a solution of the equation which satisfies

$$2\sigma\frac{\partial \Psi}{\partial \sigma} + \Psi = 0 \tag{10.4.6}$$

on the cone. Unfortunately this would imply $\Psi \propto \sigma^{-1/2}$, and then there could be no contribution from the bracket in the first term, which is evaluated at $\sigma = 0$.

Instead we look for a solution of

$$\frac{\partial^2 \Psi}{\partial x^2} + \frac{\partial^2 \Psi}{\partial y^2} - \frac{\partial^2 \Psi}{\partial z^2} = 0 \tag{10.4.7}$$

for which $\Psi = 0$ everywhere on Γ.

Consider solutions of the form

$$\Psi(x, y, z) = \Psi(\zeta),$$

where $\{(x - x_0)^2 + (y - y_0)^2\}\zeta^2 = (z - z_0)^2$. For convenience let us write for the moment

$$\rho^2 = (x - x_0)^2 + (y - y_0)^2.$$

Then

$$\frac{\partial^2 \Psi}{\partial x^2} = \frac{d^2\Psi}{d\zeta^2}\left(\frac{\partial \zeta}{\partial x}\right)^2 + \frac{d\Psi}{d\zeta}\frac{\partial^2 \zeta}{\partial x^2}$$

and

$$\frac{\partial \zeta}{\partial x} = -\frac{\zeta}{\rho}\frac{\partial \rho}{\partial x} = -\frac{\zeta}{\rho^2}(x - x_0), \frac{\partial^2 \zeta}{\partial x^2} = \zeta\{3(x - x_0)^2 - \rho^2\}\rho^4$$

with similar expressions for $\partial\zeta/\partial y$ and $\partial^2\zeta/\partial y^2$, while

$$\frac{\partial \zeta}{\partial z} = \frac{1}{\rho} \quad \text{and} \quad \frac{\partial^2 \zeta}{\partial z^2} = 0.$$

Hence

$$\frac{1}{\rho^2}\left\{(\zeta^2 - 1)\frac{d^2\Psi}{d\zeta^2} + \zeta\frac{d\Psi}{d\zeta}\right\} = 0. \tag{10.4.8}$$

A solution of this equation is

$$\Psi(\zeta) = \cosh^{-1}\zeta = \log\{\zeta + (\zeta^2 - 1)^{1/2}\} \tag{10.4.9}$$

and this vanishes identically on the cone Γ. There is unfortunately a

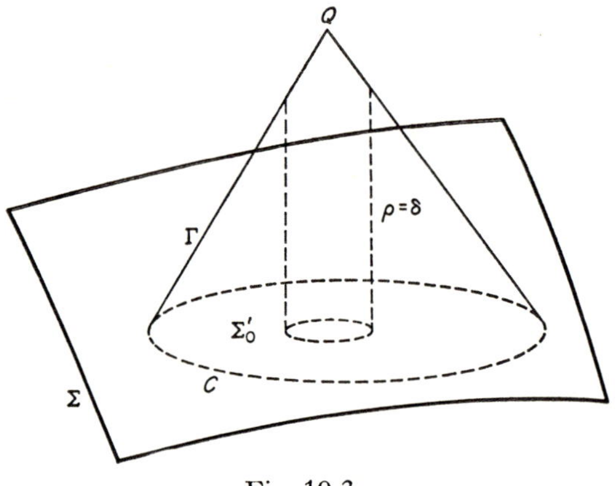

Fig. 10.3

new difficulty introduced, since we now have a singularity at every point of the line $\rho = 0$. Instead, therefore, of integrating over all the interior of V, we shall integrate over the region V' obtained by removing from V its common part with the cylinder $\rho = \delta$. Σ_0' is defined similarly (see Fig. 10.3).

Equation (10.4.4) is then amended to

$$\int_{V'} \cosh^{-1}\{(z - z_0)/\rho\} f(x, y, z)\, d\tau$$
$$= \int_{\Sigma_0'} \left[\frac{\partial \Phi}{\partial \nu} \cosh^{-1}\left\{\frac{z - z_0}{\rho}\right\} - \Phi \frac{\partial}{\partial \nu} \cosh^{-1}\left\{\frac{z - z_0}{\rho}\right\}\right] dS$$
$$+ \int_{\rho=\delta} \left[\frac{\partial \Phi}{\partial \nu} \cosh^{-1}\left\{\frac{z - z_0}{\rho}\right\} - \Phi \frac{\partial}{\partial \nu} \cosh^{-1}\left\{\frac{z - z_0}{\rho}\right\}\right] dS. \tag{10.4.10}$$

On the cylinder

$$dS = \delta\, d\theta\, dz,$$

and so the first term in the last integral is of order $\delta \log \delta$, and vanishes in the limit as $\delta \to 0$. On the other hand the conormal and normal coincide on the cylinder, since all normals to the surface are orthogonal to the z-axis and so on the cylinder

$$\frac{\partial}{\partial \nu} \cosh^{-1}\left\{\frac{z - z_0}{\rho}\right\} = -\frac{\partial}{\partial \rho} \cosh^{-1}\left\{\frac{z - z_0}{\rho}\right\} = +\frac{z - z_0}{\rho\{(z - z_0)^2 - \rho^2\}^{1/2}}.$$

Finally then

$$\int \Phi \frac{\partial}{\partial \nu} \cosh^{-1}\left(\frac{z - z_0}{\rho}\right) dS$$
$$= \int_0^{2\pi} d\theta \int_{z_1'(\theta)}^{z_2'(\theta)} \frac{z - z_0}{\{(z - z_0)^2 - \delta^2\}^{1/2}} \Phi(x_0 + \delta \cos\theta, y_0 + \delta \sin\theta, z)\, dz \tag{10.4.11}$$

where the z-coordinate is $z_1'(\theta)$ where the cylinder cuts Σ, and $z_2'(\theta)$ where it cuts the cone Γ. In the limit $\delta \to 0$, let $z_1'(\theta) \to z_1$ and then $z_2'(\theta) \to z_0$ so that the limit of the right-hand side of (10.4.11) is

$$2\pi \int_{z_1}^{z_0} \Phi(x_0, y_0, z)\, dz. \tag{10.4.12}$$

From (10.4.10) then:

$$2\pi \int_{z_1}^{z_0} \Phi(x_0, y_0, z)\, dz = -\int_V \cosh^{-1}\left\{\frac{z - z_0}{\rho}\right\} f(x, y, z)\, d\tau$$
$$+ \int_{\Sigma_0} \left[\frac{\partial \Phi}{\partial \nu} \cosh^{-1}\left\{\frac{z - z_0}{\rho}\right\} - \Phi \frac{\partial}{\partial \nu} \cosh^{-1}\left\{\frac{z - z_0}{\rho}\right\}\right] dS. \tag{10.4.13}$$

We obtain an integral involving $\Phi(x, y, z)$, as was suggested in the previous section. The desired answer can be recovered quite simply from this, by differentiating with respect to the coordinate z_0. The final solution of the boundary value problem is:

$$2\pi\Phi\{x_0, y_0, z_0\} = \frac{\partial}{\partial z_0}\int_{\Sigma_0}\left[\frac{\partial\Phi}{\partial\nu}\cosh^{-1}\left\{\frac{z-z_0}{\rho}\right\}\right.$$
$$\left. - \Phi\frac{\partial}{\partial\nu}\cosh^{-1}\left\{\frac{z-z_0}{\rho}\right\}\right] dS - \frac{\partial}{\partial z_0}\int_V \cosh^{-1}\left\{\frac{z-z_0}{\rho}\right\} f(x, y, z)\, d\tau. \tag{10.4.14}$$

10.5 Hadamard finite values of infinite integrals

It was observed by HADAMARD [8] that the formula (10.4.14) has points of considerable interest. If we take first the term:

$$\frac{\partial}{\partial z_0}\int_{\Sigma_0}\frac{\partial\Phi}{\partial\nu}\cosh^{-1}\left\{\frac{z-z_0}{\rho}\right\} dS, \tag{10.5.1}$$

the coordinate z_0 does not appear in $\partial\Phi/\partial\nu$, so that provided it is sufficiently well-behaved to differentiate under the integral sign, (10.5.1) can be written alternatively as:

$$\int_{\Sigma_0}\frac{\partial\Phi}{\partial\nu}\frac{dS}{\{(z-z_0)^2-\rho^2\}^{1/2}}. \tag{10.5.2}$$

Now $\{(z-z_0)^2-\rho^2\}^{-1/2}$ is a solution of the wave equation (10.4.7), just as $\{(z-z_0)^2+\rho^2\}^{-1/2}$ is a solution of the three-dimensional Laplace equation, and so (10.5.2) looks like a direct analogy with the corresponding term in formula (6.4.4) (with $n = 3$), which we have given for the Laplace equation. There is one very great difference however; whereas $\{(z-z_0)^2+\rho^2\}^{+1/2}$ vanishes only at the point (x_0, y_0, z_0), $\{(z-z_0)^2-\rho^2\}^{1/2}$ vanishes at every point on the surface of the cone Γ. It so happens that the intersection of Γ and Σ_0 is the curve C, and so the singularity of the integrand in (10.5.2) is an integrable one, and we may therefore use (10.5.2).

The difficulties raised by this difference become quite real when we go to the second term:

$$\frac{\partial}{\partial z_0}\int_{\Sigma_0}\Phi\frac{\partial}{\partial\nu}\cosh^{-1}\{(z-z_0)/\rho\}\, dS. \tag{10.5.3}$$

If we proceed as above, we shall rewrite this as:

$$\int_{\Sigma_0}\Phi\left[\frac{\partial}{\partial\nu}\{(z-z_0)^2-\rho^2\}^{-1/2}\right] dS, \tag{10.5.4}$$

and this is divergent. To be more strict, if we replace Σ_0 by $\Sigma_0{}^*$ which is a region of Σ contained within Σ_0, and whose boundary C^* is everywhere a distance δ from the boundary C of Σ_0, then the integral (10.5.4) is of order $\delta^{-1/2}$.

It is therefore roughly true to say that, for most practical purposes in evaluating (10.5.1), it is of little consequence in which order the differentiation and the integration are performed, the answer obtained will be identical. It is usually more convenient to perform the differentiation first and so to replace (10.5.1) by (10.5.2). In the case of (10.5.3), however, the integration must be performed first, otherwise we are faced with an infinity.

Hadamard further pointed out, however, that by the adoption of quite a simple device, even the term (10.5.3) could be treated in direct analogy with the corresponding solution of the Laplace equation.

Let $[a, b]$ be a closed interval of the real axis, and let $f(x)$ have a continuous first derivative on $[a, b]$; also let ν be a constant in the range $0 < \nu < 1$, and ε be small and positive, then if $a < x < b$,

$$\int_a^{x-\varepsilon} f(\xi)\,(x-\xi)^{-1-\nu}\,d\xi - g(x)\int_a^{x-\varepsilon}(x-\xi)^{-1-\nu}\,d\xi, \qquad (10.5.5)$$

will diverge as $\varepsilon \to 0$ almost everywhere on $[a, b]$ unless

$$g(x) = f(x). \qquad (10.5.6)$$

In this case we may say, with temporary disregard for mathematical niceties, that

$$\lim_{\varepsilon\to 0}\int_a^{x-\varepsilon} f(\xi)\,(x-\xi)^{-1-\nu}\,d\xi$$
$$= \lim_{\varepsilon\to\infty}\left[\int_a^{x-\varepsilon}\{f(\xi)-f(x)\}(x-\xi)^{-1-\nu}\,d\xi + f(x)\int_a^{x-\varepsilon}(x-\xi)^{-1-\nu}\,d\xi\right],$$

or since the first limit on the right is finite and well behaved, we may integrate by parts and so

$$\lim_{\varepsilon\to 0}\nu\int_a^{x-\varepsilon} f(\xi)\,(x-\xi)^{-1-\nu}\,d\xi$$
$$= -f(a)\,(x-a)^{-\nu} - \int_a^{x} f'(\xi)\,(x-\xi)^{-\nu}\,d\xi + \lim_{\varepsilon\to 0} f(x)\,\varepsilon^{-\nu}.$$

Of the terms on the right, only the last becomes infinite, and this is quite uniquely defined. We may then, without ambiguity, define 'the finite part' of a divergent integral of this kind, for which we use the symbol

$$*\!\!\int.$$

From the formula above then

$$\nu^* \int_a^x f(\xi)\,(x-\xi)^{-1-\nu}\,d\xi = -f(a)\,(x-a)^{-\nu} - \int_a^x f'(\xi)\,\{x-\xi\}^{-\nu}\,d\xi. \tag{10.5.7}$$

We may generalize this a little. We now suppose that $f(x)$ has its first k derivatives continuous on $[a, b]$. Then

$$\lim_{\varepsilon\to 0} (\nu+k-1)(\nu+k-2)\ldots\nu\int_a^{x-\varepsilon} f(\xi)\,(x-\xi)^{-(k+\nu)}\,d\xi$$
$$= (\nu+k-1)(\nu+k-2)\ldots$$
$$\times\,\nu\int_a^{x-\varepsilon}\left[f(\xi)-\sum_{r=0}^{k-1}\frac{f^{(r)}(x)}{r!}(\xi-x)^r\right](x-\xi)^{-\nu-k}\,d\xi$$
$$+ (\nu+k-1)(\nu+k-2)\ldots$$
$$\times\,\nu\sum_{r=0}^{k-1}(-1)^r\frac{f^{(r)}x}{r!}\int_a^{x-\xi}(x-\xi)^{r-\nu-k}\,d\xi$$

and on performing k integrations by parts:

$$\lim_{\varepsilon\to 0} (\nu+k-1)(\nu+k-2)\ldots\nu\int_a^{x-\varepsilon} f(\xi)\,(x-\xi)^{-(k+\nu)}\,d\xi$$
$$= -(x-a)^{1-k-\nu}\left\{\sum_{r=0}^{k-1}(-1)^r(\nu+k-r-2)\right.$$
$$\left.\times\,(\nu+k-r-3)\ldots\nu f^{(r)}(a)\,(x-a)^r\right\}$$
$$+ (-1)^k\int_a^x f^{(k)}(\xi)\,(x-\xi)^{-\nu}\,d\xi + (\nu+k-1)$$
$$\times\,(\nu+k-2)\ldots\nu\lim_{\varepsilon\to 0}\sum_{r=0}^{k-1}(-1)^r\frac{f^{(r)}(x)}{r!}\,\frac{\varepsilon^{r+1-\nu-k}}{\nu+k-r-1}.$$

An obvious extension of (10.5.7) is therefore

$$(\nu+k-1)(\nu+k-2)\ldots\nu^*\int_a^x f(\xi)\,(x-\xi)^{-k-\nu}\,d\xi$$
$$= -(x-a)^{1-k-\nu}\sum_{r=0}^{k-1}(-1)^r\frac{\Gamma(\nu+k-r-1)}{\Gamma(\nu)}(x-a)^r f^{(r)}(a)$$
$$+ (-1)^k\int_a^x f^{(k)}(\xi)\,(x-\xi)^{-\nu}\,d\xi. \tag{10.5.8}$$

It should be noted that the uniqueness of the definition of the finite part of the integrals considered here arises because, if $0 < \nu < 1$, and k is an integer positive or negative, then

$$\lim_{\varepsilon\to +0} \varepsilon^{k-\nu}$$

is either zero or infinity depending on the sign of k, and so if the terms becoming infinite at the limit are removed, no other terms at the limit contribute to the integral. This would not be true if we let $\nu = 1$, and so there is no analogous definition of

$$*\int_a^x f(\xi)\,(x-\xi)^{-k}\,d\xi$$

for integer k.

We may further extend our definition to multiple integrals. Suppose that V is a region of n-dimensional space and its boundary Σ can be considered composed of a piece of hypersurface Σ_0, which we assume is regular, the rest being Σ'. Further, let d be the shortest distance from a point inside V to the hypersurface Σ_0. If, as before, $0 < \nu < 1$, and k is a positive integer or zero,

$$\int_V F(\mathbf{x})\,d^{-\nu-k-n}\,d\tau \tag{10.5.9}$$

diverges, where $F(\mathbf{x})$ is assumed to have continuous partial derivatives of all orders $\leqslant k$.

Now let $V = V_0 + V'$, where V_0 is defined by

$$d \leqslant \delta$$

for a suitable choice of δ. If δ is small enough we can choose in V_0 a set of curvilinear coordinates $\eta_1, \eta_2, \ldots, \eta_n$, so that the surface Σ_0 is $\eta_n = 0$, and the surface generated by $d = \delta$ is $\eta_n = \delta$ (this can be done with suitable scaling). Then we can define

$$*\int_V F(\mathbf{x})\,d^{-\nu-k-n}\,d\tau = \int_{V'} F(\mathbf{x})\,d^{-\nu-k-n}\,d\tau + \int d\eta_1\,d\eta_2\ldots d\eta_{n-1}$$
$$\times *\int_0^\delta \mathcal{J}F(\mathbf{x})\,d^{-\nu-k-n}\,d\eta_n, \tag{10.5.10}$$

where $\mathcal{J}$ is the appropriate Jacobian.

Before we apply these ideas to the solution of the wave equation, we wish to look at the simplest properties of the 'finite' parts of the divergent integrals.

In a number of simple ways these parts behave like ordinary integrals, for example:

$$*\int_a^x f(\xi)\,(x-\xi)^{-\nu-k}\,d\xi = \int_a^c f(\xi)\,(x-\xi)^{-\nu-k} + *\int_c^x f(\xi)\,(x-\xi)^{-\nu-k}\,d\xi,$$

where $a < c < x < b$, and

$$*\int_a^x \{f(\xi) + g(\xi)\}(x-\xi)^{-\nu-k}\,d\xi$$
$$= *\int_a^x f(\xi)\,(x-\xi)^{-\nu-k}\,d\xi + *\int_a^x g(\xi)\,(x-\xi)^{-\nu-k}\,d\xi.$$

On the other hand, inequalities which are satisfied by ordinary integrals will not, in general, be satisfied by Hadamard integrals, for example

$$(x-\xi)^{-\nu-1} > 0, \quad a < \xi < x,$$

yet

$$*\int_a^x (x-\xi)^{-\nu-1}\, d\xi = -(x-a)^{-\nu} < 0.$$

Let $f(x)$ have a continuous first derivative as in (10.5.5), and as usual let $0 < \nu < 1$:

$$F(x) = \int_a^x f(\xi)\,(x-\xi)^{-\nu}\, d\xi. \tag{10.5.11}$$

Then

$$F'(x) = \lim_{\varepsilon\to 0} \frac{d}{dx}\int_0^{x-\varepsilon} f(\xi)\,(x-\xi)^{-\nu}\, d\xi,$$

$$= \lim_{\varepsilon\to 0}\left\{-\nu\int_a^{x-\varepsilon} f(\xi)\,(x-\xi)^{-\nu-1}\, d\xi + f(x-\varepsilon)\varepsilon^{-\nu}\right\},$$

and when we integrate the first term by parts

$$F'(x) = f(a)\,(x-a)^{-\nu} + \int_a^x f'(\xi)\,(x-\xi)^{-\nu}\, d\xi,$$

$$= -\nu{*}\int_a^x f(\xi)\,(x-\xi)^{-1-\nu}\, d\xi.$$

In other words

$$F'(x) = *\int_a^x f(\xi)\frac{d}{dx}(x-\xi)^{-\nu}\, d\xi. \tag{10.5.12}$$

We see also that if $k \geqslant 1$ and $f(x)$ possesses $(k+1)$ continuous derivatives on $[a, b]$

$$\frac{d}{dx}\left\{\int_a^{x-\varepsilon} f(\xi)\,(x-\xi)^{-\nu-k}\, d\xi + \sum_{r=0}^{k-1}(-1)^r \frac{f^{(r)}(x)}{r!(r+1-\nu-k)}\varepsilon^{r+1-\nu-k}\right\}$$

$$= -(\nu+k)\int_a^{x-\varepsilon} f(\xi)\,(x-\xi)^{-\nu-k+1}\, d\xi + f(x-\varepsilon)\varepsilon^{-\nu-k}$$

$$+ \sum_{r=0}^{k-1}(-1)^r \frac{f^{(r+1)}(x)}{r!(r+1-\nu-k)}\varepsilon^{r+1-\nu-k},$$

$$= -(\nu+k)\left\{\int_a^{x=\varepsilon} f(\xi)\,(x-\xi)^{-\nu-k+1} + \sum_{r=0}^{k}(-1)^r \frac{f^{(r)}(x)}{r!(r-\nu-k)}\varepsilon^{r-\nu-k}\right\}$$

$$+ (-1)^{k+1}\frac{f^{(k+1)}(\tilde{x})\,\varepsilon^{1-\nu}}{(k+1)!},$$

where $\bar{x}$ is some average value of x. Thus:

$$\frac{d}{dx}{}^*\int_a^x f(\xi)\,(x-\xi)^{-\nu-k}\,d\xi = -(\nu+k)^*\int_a^x f(\xi)\,(x-\xi)^{-\nu-k-1}\,d\xi. \tag{10.5.13}$$

Provided therefore that $f(x)$ is suitably well behaved, we may differentiate Hadamard type integrals, under the integral sign. We can also differentiate ordinary integrals under the integral sign and, when appropriate, interpret the answer so obtained as a Hadamard integral.

All this is very much less surprising today than it seemed when Hadamard first derived these results. What we can now perceive, in the light of recent work on the subject, is that we are dealing here with generalized functions or distributions. The derivatives obtained above are also distribution derivatives, and these are shown to exist to all orders. This work on generalized functions has very great implications for the theory of linear partial differential equations. An elementary introduction to the theory of distributions is given in Chapter 11. For a fuller discussion of the properties of Hadamard type integrals, and in particular of the properties of multiple integrals, the reader is referred to Hadamard's own words [8].

We may now return to the discussion of the terms (10.5.1) and (10.5.2), and observe that the correct interpretation of (10.5.4) is as an integral of Hadamard type. Thus the formula which would be analogous to (6.4.4) with $n = 3$ is, from (10.4.14):

$$2\pi\Phi(x_0, y_0, z_0) = \int_{\Sigma_0} \frac{\partial \Phi}{\partial \nu}\left(\frac{1}{\gamma}\right) dS - {}^*\int_{\Sigma_0} \Phi \frac{\partial}{\partial \nu}\left(\frac{1}{\gamma}\right) dS - \frac{\partial}{\partial z_0}\int_V \cosh^{-1}\{(z-z_0)/\rho\}\, f(x, y, z)\, d\tau, \tag{10.5.14}$$

where we have put

$$\gamma^2 = (z-z_0)^2 - (x-x_0)^2 - (y-y_0)^2 = (z-z_0)^2 - \rho^2.$$

It should finally be remarked that although we have assumed, in the analysis above, that the conoid Γ and the surface Σ intersect in a closed curve C, there is no necessity for this. They are just as likely to intersect in an open curve, provided always that the integrals over the region of surface cut off still converge. This would occur if, for example, Σ were a plane parallel to the axes of the characteristic cone.

10.6 The wave equation in more than three variables

We may try to extend the work of the previous section to finding a solution of the equation:

$$\frac{\partial^2 \Phi}{\partial z^2} - \nabla_n{}^2 \Phi = f(\mathbf{x}, z), \tag{10.6.1}$$

where as usual

$$\nabla_n{}^2\Phi = \frac{\partial^2\Phi}{\partial x_i\,\partial x_i}.$$

This equation is self-adjoint and of normal hyperbolic type, and the characteristic cone and conoid coincide. The cone Γ with vertex at the point Q with coordinates $(x_{10}, \ldots x_{n0}, z_0)$ or $(\mathbf{x}_0, z_0)$ is:

$$\gamma^2 \equiv (z - z_0)^2 - |\,\mathbf{x} - \mathbf{x}_0\,|^2 = (z - z_0)^2 - \rho^2 = 0. \qquad (10.6.2)$$

As in the previous section we shall assume that we are given the values of Φ and $\partial\Phi/\partial n$ on a surface Σ, which is piecewise regular, and nowhere characteristic. We shall further suppose that Γ and Σ intersect in a hypercurve C, which encloses a region Σ_0 of Σ. The volume of space between Γ and Σ we shall call V, as before.

By direct substitution into the homogeneous equivalent of (10.6.1) we can see that the 'fundamental' solution, in analogy with that for the $(n + 1)$-dimensional Laplace equation in Section 6.4, is:

$$\Phi = \gamma^{1-n}. \qquad (10.6.3)$$

If n is odd, say $2m + 1$, then

$$\Phi = \{(z - z_0)^2 - \rho^2\}^{-m}$$

and the singularity on the characteristic cone is of integer order; we have shown above that no unique definition of the finite part of a divergent integral is possible in this case.

First, then, we shall consider the case n even, equal to $2m$, say, when the singularity is of order half an odd integer, and we can apply the foregoing theory. There still remains one minor difficulty. Since every point of the cone surface is a singularity, the vertex of the cone, which is a point of confluence of the surface, is a singularity of higher order. This difficulty we can deal with in the usual way. We derive the region V' from V by excluding from V its intersection with the hypersphere $(z - z_0)^2 + \rho^2 = \delta^2$. Let S_δ be the part of the spherical surface cut off by Γ. Then

$$\begin{aligned}\int_{V'} f(\mathbf{x}, z)\gamma^{1-2m}\,d\tau = {}^*\!\!\int_{\Sigma_0}\left\{\frac{\partial\Phi}{\partial\nu}\gamma^{1-2m} - \Phi\frac{\partial}{\partial\nu}(\gamma^{1-2m})\right\}dS \\ + {}^*\!\!\int_{\Gamma}\left\{\frac{\partial\Phi}{\partial\nu}\gamma^{1-2m} - \Phi\frac{\partial}{\partial\nu}(\gamma^{1-2m})\right\}dS \\ + {}^*\!\!\int_{S_\delta}\left\{\frac{\partial\Phi}{\partial\nu}\gamma^{1-2m} - \Phi\frac{\partial}{\partial\nu}(\gamma^{1-2m})\right\}dS. \qquad (10.6.4)\end{aligned}$$

The second integral on the right-hand side vanishes identically. On the surface of the sphere S_δ, $dS = \delta^{2m}\,d\Omega_{n+1}$ where Ω_{n+1} is the surface area of the unit sphere in $(n + 1)$ dimensions (see Appendix 2), while γ^{1-2m}

is $O(\delta^{1-2m})$ away from the cone Γ. Therefore there will be no contribution, either, from the first term in the last integral.

The second term in the last integral will contribute

$$\overline{\Phi}^* \int_{S_\delta} \frac{\partial}{\partial \nu}(\gamma^{1-2m})\, dS,$$

where $\overline{\Phi}$ is an average value taken on the surface.

Now

$$\text{grad}_{n+1}\,\gamma = \left(-\frac{\mathbf{x}-\mathbf{x}_0}{\gamma}, \frac{z-z_0}{\gamma}\right)$$

and

$$\boldsymbol{\nu} = \left(-\frac{\mathbf{x}-\mathbf{x}_0}{r}, \frac{z-z_0}{r}\right)$$

where $r^2 = (z-z_0)^2 + \rho^2$, so that

$$\frac{\partial}{\partial \nu}(\gamma^{1-2m}) = -(2m-1)\frac{r}{\gamma^{2m+1}}.$$

This integral is therefore equivalent to:

$$-(2m-1)\delta^2 {}^* \int (\delta^2 - \rho^2)^{-1/2}(\delta^2 - 2\rho^2)^{-m-1/2}\, dX$$

where

$$dX = dx_1\, dx_2 \ldots dx_{2m},$$

and the integral is to be taken over the interior of the n-dimensional sphere

$$2\rho^2 = 2\,|\mathbf{x}-\mathbf{x}_0|^2 \leqslant \delta^2,$$

since this inequality is satisfied by all points on S_δ, being within the cone Γ.

This, if we put $\rho = (\delta/\sqrt{2})\xi$, can be written alternatively as

$$-(2m-1)2^{-m}\Omega_m {}^* \int_0^1 \left(1 - \frac{\xi^2}{2}\right)^{-1/2} (1-\xi^2)^{-m-1/2}\,\xi^{2m-1}\, d\xi \qquad (10.6.5)$$

or, replacing ξ^2 by η,

$$(2m-1)2^{-m-1/2}\Omega_{2m}{}^* \int_0^1 \frac{\eta^{m-1}d\eta}{(2-\eta)^{1/2}(1-\eta)^{m+1/2}} = -A_{2m}\Omega_{2m}, \text{ say.}$$

From (10.5.8), if $(2-\eta)^{1/2} g(\eta) = \eta^{m-1}$

$$\Gamma(m+\tfrac{1}{2}) {}^* \int_0^1 \frac{\eta^{m-1}\, d\eta}{(2-\eta)^{1/2}(1-\eta)^{m+1/2}} = \Gamma(m+\tfrac{1}{2})^* \int_0^1 \frac{g(\eta)\, d\eta}{(1-\eta)^{m+1/2}}$$

$$= -\sum_{r=0}^{m-1} (-1)^r \Gamma(m-r-\tfrac{1}{2}) g^r(0)$$

$$+ (-1)^m \sqrt{\pi} \int_0^1 g^{(m)}(\eta)\,(1-\eta)^{-1/2}\, d\eta. \qquad (10.6.6)$$

We may therefore substitute $-A_{2m}\Omega_{2m}$ for the integral (10.6.5), where the constant A_{2m} is defined by (10.6.6). It can be verified that $A_2 = 1$ in accordance with the result of the previous section.

In the limit then as $\delta \to 0$

$$\int_\Omega f(\mathbf{x}, z)\gamma^{1-2m}\, d\tau = {}^*\!\int_{\Sigma_0}\left\{\frac{\partial \Phi}{\partial \nu}\gamma^{1-2m} - \Phi\frac{\partial}{\partial \nu}(\gamma^{1-2m})\right\} dS - A_{2m}\Omega_{2m}\,\Phi(x_0, z_0). \tag{10.6.7}$$

This is substantially the result we obtained above when $m = 1$, only now the first term on the right-hand side must also be a Hadamard integral.

In the expression (10.6.6) it will be noticed that, in fact,

$$g(0) = g'(0) = \ldots = g^{(m-2)}(0) = 0,$$

but $g^{(m-1)}(0) \neq 0$. Thus the right-hand side of (10.6.6) reduces to:

$$(-1)^m\sqrt{\pi}\left\{g^{m-1}(0) + \int_0^1 g^{(m)}(\eta)\,(1-\eta)^{-1/2}\,d\eta\right\}.$$

From the Leibnitz formula:

$$g^{(m-1)}(0) = \frac{(m-1)!}{\sqrt{2}},$$

while

$$g^{(m)}(\eta) = \sum_{k=0}^{m-1}\binom{m}{k}\frac{\Gamma(m)\,\Gamma(m-k+\frac{1}{2})}{\Gamma(m-k)\,\Gamma(\frac{1}{2})}\eta^{m-k-1}\,(2-\eta)^{m-k+1/2}$$

and

$$\int_0^1\left(\frac{\eta}{2-\eta}\right)^{m-k-1}\frac{d\eta}{(2-\eta)^{3/2}\,(1-\eta)^{1/2}} = \frac{1}{\sqrt{2}}\int_0^1(1-\zeta)^{m-k-1}\,\zeta^{-1/2}\,d\zeta = \frac{1}{\sqrt{2}}\frac{\Gamma(m-k)\,\Gamma(\frac{1}{2})}{\Gamma(m-k+\frac{1}{2})},$$

if we use the transformation

$$(2-\eta)\zeta = 2(1-\eta).$$

Finally therefore

$$A_{2m} = \frac{(-1)^{m-1}\sqrt{\pi}\Gamma(m)}{2^m\Gamma(m-\frac{1}{2})}\left\{1 + \sum_{k=0}^{m-1}\binom{m}{k}\right\} = \frac{(-1)^m\sqrt{\pi}\Gamma(m)}{\Gamma(m-\frac{1}{2})}. \tag{10.6.8}$$

10.7 Conclusion

The corresponding result for a space with an odd number of dimensions can be deduced as a special case of a space of one dimension higher, by the method of reduction we have used before, in Chapter 7. Hadamard does this, but the analysis is exceedingly complex, and part of the reason for this arises from incorporating into the analysis the unconventional idea of the finite part of a divergent integral.

This additional complexity seems intuitively to be wrong, since both the one-dimensional and three-dimensional wave equations are included among those we are considering, and their solutions were found before the solution of the two-dimensional equation was known.

The next chapter gives a brief account of the theory of generalized functions, and shows how the theory is relevant to the study of partial differential equations. We shall be able to obtain solutions to the wave equation in an odd number of variables, by the reduction technique suggested, but avoiding some of the heaviness of Hadamard's method. Nevertheless the basic paradox is unresolved, that it is easier by these methods to solve the wave equation in an even-dimensioned space than in an odd-dimensioned one, and yet the simplest solutions known are both for odd-dimensioned spaces.

EXERCISES

10.1. Show that if the equation (10.1.1) is self-adjoint, the Riemann function desired is the solution of (10.1.1) which is constant on each of the lines $x = x_0$, $y = y_0$.

10.2. Using equation (10.4.14) find the solutions of the equation

$$\Phi_{xx} + \Phi_{yy} = \Phi_{zz}$$

in the region $z \geqslant 0$ when

(a) $$\Phi = 0, \quad \frac{\partial \Phi}{\partial z} = 1, \text{ on } z = 0.$$

(b) $$\Phi = 1, \quad \frac{\partial \Phi}{\partial z} = 0, \text{ on } z = 0.$$

10.3. Again using equation (10.4.14) find the solution of the same equation in $x > 0$ when

$$\Phi = 0 \text{ everywhere on } x = 0, \quad \frac{\partial \Phi}{\partial x} = 0 \text{ on } x = 0$$

except when $0 \leqslant z \leqslant 1$ and on this strip $\partial\Phi/\partial x = 1$.

10.4. Find the Hadamard finite values of the following:

(a) $$\int_0^1 \frac{x\,dx}{(1-x)^{3/2}}$$

(b) $$\frac{d}{dy}\int_0^y \frac{x^2\,dx}{(y-x)^{5/2}}, \quad y > 0$$

(c) $$\int_{-1}^1 \frac{dx}{(1-x^2)^{3/2}} \quad \left(\text{treat this as } \int_{-1}^0 + \int_0^1\right).$$

10.5. Verify the answers to 10.2 and 10.3 using Hadamard's formulae.

ANSWERS

10.2. (*a*) $\Phi(x_0, y_0, z_0) = z_0$, (*b*) $\Phi(x_0, y_0, z_0) = 1$ (both of course obvious without using the formulae).

10.3. A solution satisfying the boundary conditions is

$$\Phi = 0 \text{ for all } z < z_0 + x_0$$

and for $z > z_0 + x_0$

$$\Phi = \frac{1}{2\pi} \iint_A \frac{dz\, dy}{\{(z - z_0)^2 - x_0^{\,2} - (y - y_0)^2\}^{1/2}}$$

where A is the area bounded by $z = 0$, $z = 1$ and the arcs of the hyperbola $(z - z_0)^2 - (y - y_0)^2 = x_0^{\,2}$.

10.4. (*a*) -2, (*b*) $-\frac{8}{3}y^{-1/2}$, (*c*) 0.

CHAPTER 11

Generalized Functions

11.1 Introduction

The modern theory of generalized functions has sprung from the fundamental paper of Laurent SCHWARTZ [14], and the alternative name of 'theory of distributions' was given to it by him. To the mathematical physicist the most obvious indications of the need to generalize the ordinary concept of a function are the non-existence of the so-called Dirac delta function, as an ordinary function, and the non-existence of the Fourier transform of unity.

Although various mathematicians have made different approaches to the idea of generalized functions, it is possible to establish an equivalence in their results.

Before the properties of the Dirac delta function as a generalized function were realized, it was conventional, in textbooks of applied mathematics which used it, to point out that though it might not itself exist, it could be considered as the limit of a sequence of functions, all of which were as well-behaved as required. Such a sequence, for example, as

$$(\pi\varepsilon)^{-1/2} \exp \{-(x/\varepsilon)^2\},$$

consists of functions, differentiable to all orders at every point, and vanishing exponentially at infinity; and as $\varepsilon \to 0$ it exhibits the properties of the Dirac function. One approach is therefore to generalize the idea of a function to include the limits of sequences of suitably well-behaved functions. This is the approach of TEMPLE [17] and LIGHTHILL [12]. Schwartz himself made use of the idea of a functional, and this is the approach that we shall use in what follows. For an extended account of the elementary theory of generalized functions see [5], and for a fuller description of their application to the theory of linear partial differential equations see [1] and [9].

11.2 The distribution as a functional

Initially we shall concern ourselves with functions of a single real variable defined in the whole real line $-\infty < x < \infty$. The support of a function $f(x)$ is defined to be the set of all values of x at which $f(x) \neq 0$. We shall say that $f(x)$ belongs to the class C^k (or, for brevity, $f \in C^k$) if $f(x)$ possesses derivatives of orders up to $f^{(k)}(x)$ at every point on the real line, and $f \in C^\infty$, if it possesses derivatives of all orders. If $f(x)$ is integrable on every finite interval $[a, b]$ chosen from $-\infty < a < b < \infty$, we say that $f(x)$ is locally integrable, and write $f \in L$. The set of all real functions $\phi(x)$ which belong to C^∞, and also have bounded support, we call the set of 'test' functions. We shall write that $\phi \in D$, if $\phi(x)$ is a test function.

If $f \in L$ and $\phi \in D$ then we put

$$\{f, \phi\} = \int_{-\infty}^{\infty} f(x)\,\phi(x)\,dx. \tag{11.2.1}$$

Since $\phi(x)$ is of bounded support, this integral converges, and so for a given function $f(x)$ we can obtain a value of $\{f, \phi\}$ for any test function $\phi(x)$. Hence the set of values $\{f, \phi\}$ are as much a characteristic of the function $f(x)$ as are the set of values that the function takes as x varies from $-\infty$ to ∞. For a given $f(x)$ we say that a set of values $\{f, \phi\}$ constitutes a functional of ϕ, and this can be stressed by writing

$$\{f, \phi\} = F\{\phi\}. \tag{11.2.2}$$

We see therefore that, with each $f(x)$ which is locally integrable, is associated a functional. On the other hand, we can define functionals not associated with locally integrable functions. An example of such a functional is one which associates the value $\phi(0)$ with every test function $\phi(x)$. This we may define by $\delta\{\phi\}$.

In fact the functionals defined by locally integrable functions have other properties. Two are of particular importance.

(*a*) If $f \in L$, $\phi_1 \in D$ and $\phi_2 \in D$, and a and b are any pair of constants

$$\{f, a\phi_1 + b\phi_2\} = a\{f, \phi_1\} + b\{f, \phi_2\}. \tag{11.2.3}$$

We therefore say that the functional is linear.

If we have a set of test functions $\{\phi_k\}$, depending on a parameter k and as $k \to \infty$, $\phi_k^{(m)}(x) \to 0$ uniformly for every derivative $m = 0, 1, 2, 3$ etc., then we shall say that the set $\{\phi_k\}$ converges sequentially to zero.

(*b*) It is then clear that if $f \in L$ and $\{\phi_k\}$ is a set of test functions which converges sequentially to zero,

$$\{f, \phi_k\} \to 0, \tag{11.2.4}$$

as $k \to \infty$. We therefore say that the functional is sequentially continuous.

It is equally clear that the functional $\delta\{\phi\}$ mentioned above is both linear and sequentially continuous, and so neither property is associated only with the functionals obtained from locally integrable functions.

We shall introduce the set D' of all linear, sequentially continuous functionals $F\{\phi\}$. We call D' the set of generalized functions or distributions.

From what we have said above, if $f \in L$, then

$$F\{\phi\} = \{f, \phi\} \in D', \tag{11.2.5}$$

but also

$$\delta\{\phi\} \in D'. \tag{11.2.6}$$

A distribution obtained from a locally integrable function $f(x)$ we call a regular distribution. Hence $\delta\{\phi\}$, which we may now designate the Dirac delta function, is a distribution which is not regular.

11.3 Some simple properties of generalized functions

We can obtain properties of distributions that are desired by extrapolating by definition from the properties of locally integrable functions. By this means no contradiction can be introduced. Some of the properties we shall need are given below.

If $f \in L$, and if $a(x)$ is another function such that $a \in C^\infty$ then

$$\{af, \phi\} = \int_{-\infty}^{\infty} f(x)\, a(x)\, \phi(x)\, dx = \{f, a\phi\}. \tag{11.3.1}$$

Hence by analogy we can define the distribution aF, if $a \in C^\infty$, by

$$aF\{\phi\} = F\{a\phi\}. \tag{11.3.2}$$

The shift operator E_h is defined by

$$E_h . f(x) = f(x+h). \tag{11.3.3}$$

Then if $f \in L$

$$\{E_{-h}.f, \phi\} = \int_{-\infty}^{\infty} f(x-h)\, \phi(x)\, dx = \int_{-\infty}^{\infty} f(x)\, \phi(x+h)\, dx$$
$$= \{f, E_h . \phi\}. \tag{11.3.4}$$

This suggests that we can give a consistent definition of the generalized function $E_{-h}.F$ by

$$E_{-h}.F\{\phi\} = F\{E_h . \phi\}. \tag{11.3.5}$$

Another property which it is difficult to express conveniently within this notation is given by:

$$\int_{-\infty}^{\infty} f(ax)\, \phi(x)\, dx = \frac{1}{a}\int_{-\infty}^{\infty} f(x)\, \phi\left(\frac{x}{a}\right) dx. \tag{11.3.6}$$

Finally a most important property is that if $f' \in L$, then

$$\int_{-\infty}^{\infty} f'(x)\, \phi(x)\, dx = -\int_{-\infty}^{\infty} f(x)\, \phi'(x)\, dx. \tag{11.3.7}$$

Hence we can define the distribution derivative F' by the relation

$$F'\{\phi\} = -F\{\phi'\}. \tag{11.3.8}$$

Since we have defined a generalized function through the idea of a functional, there is no question of associating a set of function values with the distribution. We can, however, say that a generalized function $F = 0$, on an open set S, if $F\{\phi\} = 0$, for every test function $\phi(x)$ whose support is contained in the complement of S. Thus, for example, we can say that the distribution $\delta = 0$, on every open set which does not contain the origin.

We may then define the support of a generalized function as being the complement of the union of all sets S on which $F = 0$. It is easily seen that if F is a regular distribution then the support of F coincides with the support of the locally integrable function $f(x)$ associated with it.

11.4 The convergence of a sequence of generalized functions

We shall say that a sequence of distributions $\{F_k\}$ converges to the functional F in D' as $k \to \infty$, if for every $\phi \in D$, $F_k\{\phi\}$ converges to $F\{\phi\}$ as $k \to \infty$. It is obvious that, if the limit exists, it is unique and it has the property of linearity. It is not obvious that the limit is sequentially continuous, and so until this is established we are not entitled to assume that the limit F is itself a distribution.

We may, however, observe two things immediately. The first is that if a sequence of locally integrable functions $\{f_k(x)\}$ converges to $f(x)$ as $k \to \infty$, and if $f \in L$ also, then $\{F_k\}$ converges to F as a regular distribution. The second is that a set of locally integrable functions $\{f_k(x)\}$ can converge to a distribution, even if it does not converge to a locally integrable function.

Simple examples of this are:

(*a*) The sequence defined by:

$$f_k(x) = k \quad \text{when} \quad 2k\,|\,x\,| \leqslant 1,$$

and

$$f_k(x) = 0 \quad \text{when} \quad 2k\,|\,x\,| > 1,$$

does not converge to any locally integrable function, but as a distribution

$$F_k \to \delta$$

as $k \to \infty$.

(*b*) The sequence $\{\sin kx\}$ does not converge as a set of locally integrable functions, but by the Riemann-Lebesgue lemma

$$\lim_{k\to\infty} \int_{-\infty}^{\infty} \phi(x) \sin kx\, dx = 0, \tag{11.4.1}$$

for every test function $\phi(x)$, and so, as a distribution, $\{\sin kx\} \to 0$ as $k \to \infty$.

Since we can integrate by parts expressions like (11.4.1) as many times as we like, we may show similarly that, considered as distributions, the sequence $\{k^\nu \sin kx\} \to 0$ as $k \to \infty$ for every constant ν. This may be used to simplify considerably the statement of the Fourier theorem.

We must, however, prove the initial theorem, which is that every convergent sequence of distributions converges to a distribution. As we said above, it is only necessary to prove that the limit F is sequentially continuous.

Suppose the contrary, so that we can find a sequence $\{\phi_r\}$ of test functions so that $\phi_r \to 0$ sequentially in D, but $F\{\phi_r\}$ does not converge to zero. Then, for some constant $c > 0$ we can find a sub-sequence $\{\phi_{r'}\}$ so that $|F\{\phi_{r'}\}| \geqslant c$. By hypothesis $\phi_{r'}$ converges sequentially to zero and so we can choose a further sub-sequence $\phi_{r''}$ so that

$$\left|\frac{d^k}{dx^k}\phi_{r''}\right| \leqslant 2^{-2r''} \text{ for } k = 0, 1, 2, \ldots, r'' - 1.$$

A new sequence which converges sequentially to zero is $\{\psi_{r''}\}$ (which we shall hereafter write as (ψ_r)) defined by $\psi_{r''} = 2^{r''}\phi_{r''}$. Then by the linear property

$$|F(\psi_r)| \geqslant 2^r c \to \infty \text{ as } r \to \infty.$$

We now choose a sub-sequence of the functions, $\{\psi_r^*\}$, and a sub-sequence of the generalized functions, $\{F_k^*\}$, in the manner detailed below.

Let ψ_1^* be chosen by the condition

$$|F\{\psi_1^*\}| > 1,$$

which is clearly possible. Since then $F_k \to F$, we can certainly also choose F_1^* so that

$$|F_1^*\{\psi_1^*\}| > 1.$$

Now choose ψ_2^* so that

$$|F_1^*\{\psi_2^*\}| < \tfrac{1}{2},$$

but so that

$$|F\{\psi_2^*\}| > |F\{\psi_1^*\}| + 2.$$

This is possible since F_1^* is a generalized function, and the sequence $\{\psi_r\}$ converges sequentially to zero. We may then, as above, select F_2^* so that also

$$|F_2^*\{\psi_2^*\}| > |F_2^*\{\psi_1^*\}| + 2,$$

which is possible since the sequence $\{F_k\}$ is convergent.

This process can be continued so that ψ_r^* is selected to satisfy the two conditions

$$|F_k\{\psi_r^*\}| < 2^{k-r} \text{ for } k = 1, 2, \ldots, r-1$$

and

$$|F\{\psi_r^*\}| > \sum_{s=1}^{r-1} |F\{\psi_s^*\} + r,$$

while we select F_r^* by the condition that

$$|F_r^*\{\psi_r^*\}| > \sum_{s=1}^{r-1} |F_r^*\{\psi_s^*\}| + r.$$

Now put

$$\psi = \sum_{s=1}^{\infty} \psi_s^*.$$

It is clear that ψ also converges sequentially since $|\psi_r| \leqslant 2^{-r} |\phi_r|$. Also

$$\sum_{s=r+1}^{\infty} |F_r^*\{\psi_s^*\}| < \sum_{s=r+1}^{\infty} 2^{r-s} = 1,$$

and so by the linear property

$$|F_r^*\{\psi\}| = \left|\sum_{s=0}^{r-1} F_r^*\{\psi_s\} + F_r^*\{\psi_r\} + \sum_{s=r+1}^{\infty} F_r^*\{\psi_s\}\right|,$$

$$\geqslant |F_r^*\{\psi_r\}| - \sum_{s=0}^{r-1} |F_r^*\{\psi_s^*\}| - \sum_{s=r+1}^{\infty} |F_r^*\{\psi_s^*\}|,$$

$$\geqslant r - 1 \to \infty \quad \text{as} \quad r \to \infty.$$

This is a contradiction since by the hypothesis $\{F_k\}$ converges. Thus the limit F is sequentially continuous, and so is a generalized function.

We can prove, further, not only that every convergent sequence of generalized functions converges to a generalized function, but that every generalized function can be considered as the limit of a sequence of special distributions. For example it can be expressed as the limit of a sequence of distributions with bounded support.

Let $a_k(x)$ be defined for every integer k by the requirements that $a_k \in C^\infty$, that $a_k(x) = 1$ when $|x| \leqslant k$, that $a_k(x) = 0$ when $|x| \geqslant 2k$, and in the remaining intervals $0 < a_k(x) < 1$, $k < |x| < 2k$.

Then from (11.3.2):

$$a_k F\{\phi\} = F\{a_k \phi\}. \tag{11.4.2}$$

For every $\phi \in D$, we can find a k_0 such that if $k > k_0$, the support of ϕ is contained within $[-k_0, k_0]$. Hence, if $k > k_0$:

$$F\{a_k \phi\} = F\{\phi\}. \tag{11.4.3}$$

On the other hand, the support of $a_k F$ is contained within $[-2k, 2k]$, and so the result is established.

Consider now test functions $\phi_\lambda(x) = \phi(\lambda, x)$ which depend on a parameter λ as well as on the independent variable x. Any functional will also clearly depend on λ, and so we may put

$$g(\lambda) = F\{\phi_\lambda\}. \tag{11.4.4}$$

If for every m $\partial^m\phi(\lambda, x)/\partial x^m$ is a continuous function of λ for some interval of λ, we may prove that $g(\lambda)$ is a continuous function of λ on that interval.

Since $\phi_\lambda(x)$ has bounded support, given λ_0 we can find a sequence $\{\lambda_k\} \to \lambda_0$ as $k \to \infty$, and for every m

$$\left| \frac{\partial^m\phi(\lambda_k, x)}{\partial x^m} - \frac{\partial^m\phi(\lambda_0, x)}{\partial x^m} \right| \to 0$$

as $k \to \infty$, uniformly in x. Hence $\phi(\lambda_k, x) - \phi(\lambda_0, x)$ converges sequentially to zero in D and so

$$F\{\phi(\lambda_k, x)\} \to F\{\phi(\lambda_0, x)\},$$

and $g(\lambda)$ is continuous at λ_0.

Furthermore since $g(\lambda)$ is continuous over some interval, then if $[\lambda_1, \lambda_2]$ is a sub-interval, the Riemann integral

$$\int_{\lambda_1}^{\lambda_2} g(\lambda)\, d\lambda$$

exists as the limit of a Riemann sum. By the linearity of the functional, the Riemann sum of the functionals is equal to the functional of the Riemann sum of ϕ_λ. We see from the convergence theorem that in the limit the functional of the Riemann sum of ϕ_λ converges to the functional of the corresponding integral of ϕ_λ. Hence

$$\int_{\lambda_1}^{\lambda_2} g(\lambda)\, d\lambda = F\left\{\int_{\lambda_1}^{\lambda_2} \phi_\lambda\, d\lambda\right\}. \tag{11.4.5}$$

Finally we see that if, for every m,

$$\frac{\partial^{m+1}\phi(\lambda, x)}{\partial\lambda\, \partial x^m}$$

is a continuous function of λ on some interval, then as $k \to 0$

$$\frac{\phi(\lambda + k, x) - \phi(\lambda, x)}{k} - \frac{\partial\phi(\lambda, x)}{\partial\lambda}$$

converges sequentially to zero in D, and so $dg/d\lambda$ exists and is equal to

$$\lim_{k\to 0} \frac{g(\lambda + k) - g(\lambda)}{k} = F\left\{\frac{\partial\phi(\lambda, x)}{\partial\lambda}\right\}. \tag{11.4.6}$$

It follows that if, for every m, $\partial^m \phi(\lambda, x)/\partial x^m$ is analytic considered as a function of the complex variable λ on some domain of the λ-plane, then $g(\lambda)$ is analytic on the same domain.

Let $\phi(x)$ be a test function, then if $f \in L$, define the convolution $f\star\phi$ by

$$f\star\phi = \int_{-\infty}^{\infty} f(y)\,\phi(x-y)\,dy, \tag{11.4.7}$$

so that $f\star\phi$ is a function of x, and clearly $f\star\phi \in C^\infty$.

If $a(x)$ is a particular test function

$$\begin{aligned}\{f\star a, \phi\} &= \int_{-\infty}^{\infty}\left\{\int_{-\infty}^{\infty} f(y)\,a(x-y)\,dy\right\}\phi(x)\,dx,\\ &= \int_{-\infty}^{\infty} f(x)\left\{\int_{-\infty}^{\infty} a(y-x)\,\phi(y)\,dy\right\}dx,\\ &= \{f, \{E_{-x}a, \phi\}\}.\end{aligned} \tag{11.4.8}$$

We may therefore consistently define the convolution of a distribution F and a test function a, $F\star a$, by

$$F\star a\{\phi\} = \int_{-\infty}^{\infty} F\{a(x-y)\}\,\phi(y)\,dy = F\{\{E_{-x}a, \phi\}\}. \tag{11.4.9}$$

Since

$$F\{a(x-y)\} \in C^\infty$$

$F\star a$ is a regular distribution.

With all these preliminaries we can now prove that any distribution can be thought of as a limit of regular distributions.

Let $a \in D$, as above, but in addition assume that for all x, $a(x) \geqslant 0$, and also that by suitable scaling

$$\int_{-\infty}^{\infty} a(x)\,dx = 1. \tag{11.4.10}$$

Now define

$$a_k(x) = ka(kx), \tag{11.4.11}$$

so that

$$\begin{aligned}\{a_k, \phi\} &= \int_{-\infty}^{\infty} a(y)\,\phi(y/k)\,dy \longrightarrow \phi(0)\int_{-\infty}^{\infty} a(y)\,dy,\\ &= \phi(0),\end{aligned}$$

as $k \longrightarrow \infty$. Hence

$$a_k \longrightarrow \delta \tag{11.4.12}$$

in D'. Furthermore

$$\int_{-\infty}^{\infty} a_k(y-x)\,\phi(y)\,dy = \int_{-\infty}^{\infty} a_k(y)\,\phi(x+y)\,dy \longrightarrow \phi(x)$$

as $k \longrightarrow \infty$, with a similar result for each derivative $\phi^m(x)$. Thus

$$F\star a_k \longrightarrow F \tag{11.4.13}$$

in D', and F can be expressed as the limit of the regular distributions $F\star a_k$.

This result shows the sort of equivalence which can be established between the Schwartz definition and the Temple definition of a generalized function.

11.5 The distribution derivatives

We can extend the definition of (11.3.8) for the distribution derivative, by induction. In fact for every m we may define

$$F^{(m)}\{\phi\} = (-1)^m F\{\phi^{(m)}\}. \tag{11.5.1}$$

Every distribution then possesses distribution derivatives of all orders, and in particular any locally integrable function has distribution derivatives of all orders.

A simple example is the Heaviside unit function $H(x)$, defined by

$$H(x) = 1, \quad \text{if} \quad x > 0,$$
$$H(x) = \tfrac{1}{2}, \quad \text{if} \quad x = 0.$$

and

$$H(x) = 0, \quad \text{if} \quad x < 0.$$

Certainly $H(x) \in L$, and so from (11.3.8) we define $H'(x)$ by

$$\{H', \phi\} = -\int_{-\infty}^{\infty} H(x)\, \phi'(x)\, dx = -\int_{0}^{\infty} \phi'(x)\, dx = \phi(0). \tag{11.5.2}$$

We may therefore put

$$H' = \delta. \tag{11.5.3}$$

Similarly

$$\{H^{(m)}, \phi\} = (-1)^m \int_{-\infty}^{\infty} H(x)\, \phi^{(m)}(x)\, dx = (-1)^{m-1} \phi^{(m-1)}(0). \tag{11.5.4}$$

Hence from (11.5.3) and (11.5.4) we may define the derivatives of the Dirac delta function by

$$\delta^{(m)}\{\phi\} = H^{(m+1)}\{\phi\} = (-1)^m \phi^{(m)}(0). \tag{11.5.5}$$

If $f \in L$ and $f'(x)$ exists, we can easily show that this derivative corresponds to the distribution derivative, for from (11.3.5) and the linear property

$$\frac{1}{h}[F\{\phi\} - E_{-h}F\{\phi\}] - F'\{\phi\} = F\left\{\frac{\phi - E_h\phi}{h} + \phi'\right\}. \tag{11.5.6}$$

As $h \to 0$

$$\frac{\phi - E_h\phi}{h} + \phi' \to 0$$

and so by sequential continuity of the right-hand side, as $h \to 0$

$$\frac{1}{h}(F - E_h F) \to F'. \tag{11.5.7}$$

11.6 The relation with Hadamard type integrals

If $f \in L$ and $f'(x)$ exists, and $\phi \in D$, we have

$$\begin{aligned} \int_{-\infty}^{\infty} \frac{d}{dx}\{f(x)\,H(x)\}\phi(x)\,dx &= -\int_{-\infty}^{\infty} f(x)\,H(x)\,\phi'(x)\,dx \\ &= -\int_{0}^{\infty} f(x)\,\phi'(x)\,dx \\ &= f(0)\,\phi(0) + \int_{0}^{\infty} f'(x)\,\phi(x)\,dx. \end{aligned} \tag{11.6.1}$$

Considered as distributions therefore

$$\frac{d}{dx}\{fH\} = f\delta + f'H. \tag{11.6.2}$$

The existence of distribution derivatives of all orders implies immediately that a meaning can be attached to integrals which, considered as ordinary integrals, would diverge.

For example, let $\phi \in D$, and $\nu > 0$, then considered as an ordinary integral

$$\begin{aligned} \nu\int_{-\infty}^{\infty} |\,x\,|^{\nu-1}\phi(x)\,dx &= \nu\int_{0}^{\infty} x^{\nu-1}\{\phi(x) + \phi(-\,x)\}\,dx \\ &= \int_{0}^{\infty} \frac{d}{dx}\{x^{\nu}\}\{\phi(x) + \phi(-\,x)\}\,dx \\ &= \int_{-\infty}^{\infty} \operatorname{sign} x \frac{d}{dx}\{|\,x\,|^{\nu}\}\phi(x)\,dx \\ &= -\int_{-\infty}^{\infty} \{\operatorname{sign} x\}|\,x\,|^{\nu}\phi'(x)\,dx. \end{aligned} \tag{11.6.3}$$

The final integral is still meaningful if $-1 < \nu < 0$, and the formula (11.6.3) is consistent with the differentiation rule. Thus (11.6.3) formally attaches a meaning to the left-hand integral which is generally divergent when $-1 < \nu < 0$, unless $\phi(0) = 0$.

Similarly, if $-1 < \nu < 0$, we may define

$$\begin{aligned} &\nu(\nu - 1)\ldots(\nu + 1 - m)\int_{-\infty}^{\infty} |\,x\,|^{\nu-m}\phi(x)\,dx \\ &\qquad = \int_{-\infty}^{\infty} \frac{d^m}{dx^m}\{(\operatorname{sign} x)^m\,|\,x\,|^{\nu}\}\,\phi(x)\,dx, \\ &\qquad = (-1)^m\int_{-\infty}^{\infty} (\operatorname{sign} x)^m\,|\,x\,|^{\nu}\phi^{(m)}(x)\,dx. \end{aligned} \tag{11.6.4}$$

We may also, if $\nu > -2$, define

$$\nu\int_{-\infty}^{\infty}(\text{sign } x)\,|\,x\,|^{\nu-1}\,\phi(x)\,dx = \int_{-\infty}^{\infty}\frac{d}{dx}\{|\,x\,|^{\nu}\}\,\phi(x)\,dx$$

$$= -\int_{-\infty}^{\infty}|\,x\,|^{\nu}\,\phi'(x)\,dx \quad (11.6.5)$$

and

$$\nu(\nu-1)\ldots(\nu+1-m)\int_{-\infty}^{\infty}(\text{sign } x)\,|\,x\,|^{\nu-m}\,\phi(x)\,dx$$

$$= \int_{-\infty}^{\infty}\frac{d^m}{dx^m}\{(\text{sign } x)^{m-1}\,|\,x\,|^{\nu}\}\,\phi(x)\,dx,$$

$$= (-1)^m\int_{-\infty}^{\infty}(\text{sign } x)^{m-1}\,|\,x\,|^{\nu}\,\phi^{(m)}(x)\,dx. \quad (11.6.6)$$

Let us apply these results to certain integrals. We suppose that $f(x)$ is differentiable as many times as we like in an interval which contains $[a, b]$. Then $f(x)$ can be continued to the rest of $(-\infty, \infty)$ as a test function. Further, let $-1 < \nu < 0$, then

$$\nu\int_a^b f(x)\,(b-x)^{\nu-1}\,dx$$

$$= \nu\int_{-\infty}^{\infty} f(x)\,\{H(b-x) - H(a-x)\}\,|\,b-x\,|^{\nu-1}\,dx.$$

Now from (11.6.3) and (11.6.5):

$$\nu\int_{-\infty}^{\infty} f(x)\,H(b-x)\,|\,b-x\,|^{\nu-1}\,dx$$

$$= \frac{\nu}{2}\int_{-\infty}^{\infty} f(x)\,|\,b-x\,|^{\nu-1}\,dx - \frac{\nu}{2}\int_{-\infty}^{\infty}\text{sign }(x-b)\,|\,b-x\,|^{\nu-1}\,dx$$

$$= -\tfrac{1}{2}\int_{-\infty}^{\infty} f'(x)\,\text{sign }(x-b)\,|\,b-x\,|^{\nu}\,dx + \tfrac{1}{2}\int_{-\infty}^{\infty} f'(x)\,|\,b-x\,|^{\nu}\,dx$$

$$= \int_{-\infty}^{\infty} f'(x)\,H(b-x)\,|\,b-x\,|^{\nu}\,dx.$$

On the other hand $|\,b-x\,|^{\nu-1}$ is analytic in $-\infty < x \leqslant a$ and so can be continued as a locally integrable function, and so by (11.6.2)

$$\nu\int_{-\infty}^{\infty} f(x)\,H(a-x)\,|\,b-x\,|^{\nu-1}\,dx$$

$$= -f(a)\,(b-a)^{\nu} + \int_{-\infty}^{\infty} f'(x)\,H(a-x)\,|\,b-x\,|^{\nu}\,dx.$$

Thus finally

$$\nu\int_a^b f(x)\,(b-x)^{\nu-1}\,dx = f(a)\,(b-a)^{\nu} + \int_a^b f'(x)\,(b-x)^{\nu}\,dx, \quad (11.6.7)$$

and by comparison with (10.5.7) this is seen to be the Hadamard finite value of the divergent integral on the left-hand side. It will be appreciated that this is not quite equivalent to the result obtained in the previous chapter, since more restrictive conditions are imposed on $f(x)$. In fact, it is possible (see for example [5]) to build up a theory of generalized functions in which the test functions are of class C^k, for any given k, and are of bounded support. If this class of test functions is used, the results will be more comparable with those obtained in the previous chapter.

In a similar way therefore, if $f(x)$ can be continued as a test function to the whole of $(-\infty, \infty)$, and if $-1 < \nu < 0$,

$$\nu(\nu-1)\ldots(\nu+1-m)\int_a^b f(x)\,(b-x)^{\nu-m}\,dx = \nu(\nu-1)\ldots$$
$$\times(\nu+1-m)\int_a^b f(x)\,\{H(b-x)-H(a-x)\}\,|\,b-x\,|^{\nu-m}\,dx.$$

From (11.6.4) and (11.6.6):

$$2\int_{-\infty}^{\infty} f(x)\,H(b-x)\,|\,b-x\,|^{\nu-m}\,dx$$
$$= \int_{-\infty}^{\infty} f(x)\,|\,b-x\,|^{\nu-m}\,dx - \int_{-\infty}^{\infty} \operatorname{sign}(x-b)\,f(x)\,|\,b-x\,|^{\nu-m}\,dx$$
$$= 2\{\nu(\nu-1)\ldots(\nu+1-m)\}^{-1}\int_{-\infty}^{\infty} f^{(m)}(x)\,H(b-x)\,|\,b-x\,|^{\nu}\,dx$$

while by an obvious generalization of (11.6.2)

$$\nu(\nu-1)\ldots(\nu+1-m)\int_{-\infty}^{\infty} f(x)\,H(a-x)\,(b-x)^{\nu-m}\,dx$$
$$= \nu(\nu-1)\ldots(\nu+1-m)\int_{-\infty}^{a} f(x)\,(b-x)^{\nu-m}\,dx$$
$$= -\nu(\nu-1)\ldots(\nu+2-m)\{f(a)\,(b-a)^{\nu+1-m}$$
$$+\int_{-\infty}^{a} f'(x)\,(b-x)^{\nu+1-m}\,dx\}$$
$$= -\nu(\nu-1)\ldots(\nu+2-m)\,f(a)\,(b-a)^{\nu+1-m}$$
$$-\nu(\nu-1)\ldots(\nu+3-m)\,f'(a)\,(b-a)^{\nu+2-m}\ldots$$
$$-f^{(m-1)}(a)\,(b-a)^{\nu}+\int_{-\infty}^{\infty} f^{(m)}(x)\,H(a-x)\,(b-x)^{\nu}\,dx.$$

If these two results are combined then

$$\nu(\nu-1)\ldots(\nu+1-m)\int_a^b f(x)\,(b-x)^{\nu-m}\,dx$$
$$= \nu(\nu-1)\ldots(\nu+2-m)\,f(a)\,(b-a)^{\nu+1-m}$$
$$+\nu(\nu-1)\ldots(\nu+3-m)\,f'(a)\,(b-a)^{\nu+2-m}+\ldots$$
$$+f^{(m-1)}(a)\,(b-a)^{\nu}+\int_a^b f^{(m)}(x)\,(b-x)^{\nu}\,dx \qquad (11.6.8)$$

which again corresponds to the Hadamard finite value, and is comparable to (10.5.8).

Since these integrals, regarded as distributions, give the Hadamard finite values, they possess the same properties as the Hadamard integrals. In the sense, therefore, of distribution derivatives, if $f \in L$ and $-1 < \nu < 0$,

$$\frac{d^m}{db^m}\int_a^b f(x)\,(b-x)^\nu\,dx = (-1)^m \nu(\nu-1)\ldots \times (\nu+1-m)\int_a^b f(x)\,(b-x)^{\nu-m}. \tag{11.6.9}$$

We can easily see that, similarly in the sense of distribution derivatives, if $f(x)$ is a test function which is also an even function, and if moreover $1 < \nu < 0$,

$$\frac{d}{d(x^2)}\int_{-x}^{x} f(\xi)\,(x^2-\xi^2)^\nu\,d\xi = \nu\int_{-x}^{x} f(\xi)\,(x^2-\xi^2)^{\nu-1}\,d\xi, \tag{11.6.10}$$

and in consequence also

$$\frac{d^m}{d(x^2)^m}\int_{-x}^{x} f(\xi)\,(x^2-\xi^2)^\nu\,d\xi = \nu(\nu-1)\ldots \times (\nu+1-m)\int_{-x}^{x} f(\xi)\,(x^2-\xi^2)^{\nu-m}\,d\xi. \tag{11.6.11}$$

Let us consider again the first integral on the right-hand side of (10.6.7), where we shall assume that the surface Σ is the plane $z=0$. Then Σ_0 is the interior of the $2m$-dimensional sphere

$$\rho^2 = |\mathbf{x}-\mathbf{x}_0|^2 = z_0^{\,2} \tag{11.6.12}$$

and $\partial/\partial\nu = -\partial/\partial z$. Hence the term reads

$$-{}^*\!\!\int_{\Sigma_0}\left[\left(\frac{\partial\Phi}{\partial z}\right)_{z=0}(z_0^{\,2}-\rho^2)^{1/2-m} + (1-2m)z_0(\Phi)_{z=0}(z_0^{\,2}-\rho^2)^{-1/2-m}\right]dX, \tag{11.6.13}$$

where, as in the last chapter, $dX = dx_1\,dx_2\ldots dx_n$.

If the problem then is really one of $(2m-1)$ dimensions, $(\Phi)_{z=0}$ and $(\partial\Phi/\partial z)_{z=0}$ will be independent of one of the coordinates, say x_{2m}. Then if we put

$$r^2 = \sum_{k=1}^{2m-1}(x_k - x_{0,k})^2$$

and $\xi = x_{2m} - x_{0,2m}$ and take as usual $d\Omega_{2m-1}$ to be an element of surface on the unit sphere in $2m-1$ dimensions, and lastly if we put

$h^2 = z_0{}^2 - r^2$, we may rewrite (11.6.13) as

$$-\int_{\Omega_{2m-1}} d\Omega_{2m-1}{}^* \int_0^{z_0} dr\, r^{2m-2}\left[\left(\frac{\partial\Phi}{\partial z}\right)_{z=0}{}^* \int_{-h}^{h} (h^2 - \xi^2)^{1/2-m}\, d\xi \right.$$

$$\left. + (1-2m)z_0(\Phi)_{z=0}{}^* \int_{-h}^{h} (h^2 - \xi^2)^{-1/2-m}\, d\right]. \quad (11.6.14)$$

Let

$$\Phi_0(r) = \frac{1}{\Omega_{2m-1}}\int_{\Omega_{2m-1}} \{\Phi\}_{z=0}\, d\Omega_{2m-1}$$

and similarly let

$$\Phi_1(r) = \frac{1}{\Omega_{2m-1}}\int_{\Omega_{2m-1}} \left\{\frac{\partial\Phi}{\partial z}\right\}_{z=0} d\Omega_{2m-1}$$

so that these are values averaged over the $(2m-1)$-dimensional sphere of radius r. Then from (11.6.11) we may write our term as

$$-\frac{\Omega_{2m-1}}{2}\int r^{2m-3}\left\{\frac{\Gamma(\frac{1}{2})}{\Gamma(m-\frac{1}{2})}\Phi_1(r)\frac{d^{m-1}}{d(r^2)^{m-1}}\int_{-h}^{h}(h^2-\xi^2)^{-1/2}\, d\xi \right.$$

$$\left. + \frac{(1-2m)\,\Gamma(\frac{1}{2})z_0}{\Gamma(m+\frac{1}{2})}\Phi_0(r)\frac{d^m}{d(r^2)^m}\int_{-h}^{h}(h^2-\xi^2)^{-1/2}\, d\xi\right\} d(r^2)$$

where, in the outside integral, the range of integration is over

$$0 \leqslant r^2 \leqslant z_0{}^2.$$

This then is

$$(-1)^m\frac{\pi^{m+1}}{2\{\Gamma(m-\frac{1}{2})\}^2}\int\left[\frac{d^{m-1}}{d(r^2)^{m-1}}\{r^{2m-3}\,\Phi_1(r)\}\right.$$

$$\left. + 2z_0\frac{d^m}{d(r^2)^m}\{r^{2m-3}\,\Phi_0(r)\}\right] d(r^2)$$

$$= (-1)^m\frac{\pi^{m+1}}{2\{\Gamma(m-\frac{1}{2})\}^2}\left[\frac{d^{m-2}}{d(z_0{}^2)^{m-2}}\{z_0{}^{2m-3}\,\Phi_1(z_0)\}\right.$$

$$\left. + 2z_0\frac{d^{m-1}}{d(z_0{}^2)^{m-1}}\{z_0{}^{2m-3}\,\Phi_0(z_0)\}\right]. \quad (11.6.15)$$

It follows that Huyghens' principle can be applied to a space with these dimensions.

We can in fact derive this same result by a method somewhat similar to that used in Section 7.8. The 'spherically' symmetric wave equation in this space is

$$r^{2m-2}\frac{\partial^2\Phi}{\partial z^2} = \frac{\partial}{\partial r}\left(r^{2m-2}\frac{\partial\Phi}{\partial r}\right), \quad (11.6.16)$$

or

$$\frac{\partial^2\Phi}{\partial r^2} + \frac{2m-2}{r}\frac{\partial\Phi}{\partial r} - \frac{\partial^2\Phi}{\partial z^2} = 0. \quad (11.6.17)$$

This is transformed by the characteristic transformation $2\xi = r - z$, $2\eta = r + z$, into the canonical form of a Darboux equation (see (13.5.14) and following),

$$(\xi + \eta)\frac{\partial^2 \Phi}{\partial \xi\, \partial \eta} + (m - 1)\left\{\frac{\partial \Phi}{\partial \xi} + \frac{\partial \Phi}{\partial \eta}\right\} = 0. \tag{11.6.18}$$

In this form it clearly possesses a solution

$$\begin{aligned}\Phi(\xi, \eta) &= \frac{\partial^{m-2}}{\partial \xi^{m-2}}\{f(\xi)\,(\xi + \eta)^{1-m}\}\\ &= \sum_{k=0}^{m-2}(-1)^k \frac{\Gamma(m+k-1)}{\Gamma(m-k-1)\,\Gamma(k+1)} f^{(m-2-k)}(\xi)\,(\xi+\eta)^{1-k-m},\end{aligned}$$

where $f(\xi)$ is an arbitrary function of ξ. Thus by an argument similar to that of Section 7.8, if $F(\mathbf{x})$ is another arbitrary function of the vector $\mathbf{x}$, a solution of the wave equation is:

$$\begin{aligned}\Phi(\mathbf{x}, z) &= \sum_{k=0}^{m-2}(-1)^k \frac{\Gamma(m+k-1)}{\Gamma(m-k-1)\,\Gamma(k+1)}\\ &\quad\times \int d\Omega_{2m-1} \int_0^\infty F(\mathbf{x} + r\boldsymbol{\zeta})\, f^{(m-2-k)}\{\tfrac{1}{2}(r - z)\} r^{m-k-1}\, dr\end{aligned} \tag{11.6.19}$$

where the outside integral is over the whole surface of the unit sphere in $2m - 1$ dimensions, and where $\boldsymbol{\zeta}$ is a unit vector.

If now we follow the same procedure as in Section 7.8, and put $f(\xi) = \delta(\xi)$, the Dirac delta function, and interpret the derivatives as distribution derivatives,

$$\begin{aligned}&\Phi(\mathbf{x}, z)\\ &= (-1)^m \Omega_{2m-1} \sum_{k=0}^{m-2} \frac{\Gamma(m+k+1)}{\Gamma(m-k-1)\,\Gamma(k+1)} 2^{m-2-k} \frac{d^{m-2-k}}{dz^{m-2-k}}\{z^{m-k-1}\bar{F}(z)\}\end{aligned} \tag{11.6.20}$$

where $\bar{F}(z)$ is the average value of $F\{\mathbf{x}\}$ taken over an $(m - 1)$-dimensional sphere of centre $\mathbf{x}$ and radius z, and this can be identified as a term of the same form as the first term on the right-hand side of (11.6.15), although it requires quite a lot of manipulation (see Appendix 5). The second term follows also by analogy with the analysis of Section 7.8.

11.7 Generalized functions of more than one independent variable

The idea of generalized functions can be extended by obvious definitions to spaces of more than one dimension. We define the set of test functions D_n to be functions defined on a space of n dimensions which have bounded support and are infinitely differentiable. Then the set

D_n' of distributions in n-dimensional space is the set of linear, sequentially continuous functionals of the test functions. In particular we can define the distribution δ_n with the property that

$$\delta_n\{\phi\} = \phi(\mathbf{0}). \tag{11.7.1}$$

Let $\mathbf{x} = \rho\boldsymbol{\xi}$, where $\boldsymbol{\xi}$ is a unit vector, then if $\nu > -n$, and $\phi \in D_n$, $dX = dx_1\, dx_2 \ldots dx_n$, and the integral is taken over the whole of the n-dimensional space R_n

$$\int \rho^\nu \phi(\mathbf{x})\, dX = \int_0^\infty \rho^{\nu+n-1} \psi(\rho)\, d\rho, \tag{11.7.2}$$

where

$$\psi(\rho) = \int_{\Omega_n} \phi(\rho\boldsymbol{\xi})\, d\Omega_n \tag{11.7.3}$$

the last integral taken over the surface of the unit sphere in n dimensions.

Since ϕ is infinitely differentiable and of bounded support, it is easily seen that ψ has an expansion in powers of ρ (of even powers only) if ρ is small and positive, is zero if ρ is greater than some bounded ρ_0, and is also infinitely differentiable. Then ψ can be continued as an even function into the region $\rho < 0$, and so $\psi(\rho)$ is a one-dimensional test function.

If, as usual, we define

$$\nabla^2 = \frac{\partial^2}{\partial x_k\, \partial x_k} \tag{11.7.4}$$

then if $\nu > 2 - n$

$$\nabla^2\rho^\nu = \nu\{\nu + n - 2\}\rho^{\nu-2}, \tag{11.7.5}$$

and both sides are locally integrable. Then if $\phi \in D_n$

$$\int_{R_n} (\nabla^2\rho^\nu)\, \phi(\mathbf{x})\, dX = \int_{R_n} \rho^\nu\{\nabla^2\phi\}\, dX = \lim_{\varepsilon\to 0} \int_{\rho>\varepsilon} \rho^\nu(\nabla^2\phi)\, dX. \tag{11.7.6}$$

It follows that in the sense of distributions we may with consistency define

$$\int_{R_n} (\nabla^2\rho^{2-n})\, \phi(\mathbf{x})\, dX = \int_{R_n} \rho^{2-n}\{\nabla^2\phi(\mathbf{x})\}\, dX.$$

From (11.7.2) and (11.7.3) we see that

$$\int_{R_n} \rho^{2-n}(\nabla^2\phi)\, dX = \int_0^\infty \rho^{2-n}\frac{\partial}{\partial\rho}\left\{\rho^{n-1}\frac{\partial}{\partial\rho}\int_{\Omega_n} \phi(\rho\boldsymbol{\xi})\, d\Omega_n\right\} d\rho,$$

$$= (n-2)\int_0^\infty \frac{\partial}{\partial\rho}\left\{\int_{\Omega_n} \phi(\rho\boldsymbol{\xi})\, d\Omega_n\right\} d\rho = (2-n)\, \Omega_n\phi(\mathbf{0}).$$

Finally therefore we see that

$$\int_{R_n} (\nabla^2\rho^{2-n})\, \phi(\mathbf{x}) dX = (2-n)\Omega_n\phi(\mathbf{0}), \tag{11.7.7}$$

and so in the sense of distributions we may extend (11.7.5) to give

$$\nabla^2 \rho^{2-n} = (2-n)\Omega_n \delta_n. \tag{11.7.8}$$

We may again rewrite (11.7.2) in the form that if $\mathscr{R}(\nu) > -n$

$$\int_{R_n} \rho^\nu \phi(\mathbf{x})\, dX = \int_0^\infty \rho^{\nu+n-1}\left[\psi(\rho) - \sum_{k=0}^{m-1} \frac{\psi^{(2k)}(0)\rho^{2k} H(1-\rho)}{\{(2k)!\}}\right] d\rho + \sum_{k=0}^{m-1} \frac{\psi^{(2k)}(0)}{\{(2k)!\}(2k+\nu+n)}. \tag{11.7.9}$$

Considered as a function of the complex variable ν, the right-hand side of (11.7.9) is regular when $\nu + n + 2m > 0$, except for simple poles at $\nu = -n, -n-2, \ldots, -n-2m+2$.

We may therefore regard this formula as giving the analytic continuation of the left-hand side into the region $\mathscr{R}(\nu) > -n - 2m$. This defines ρ^ν for all ν as a distribution with simple poles at $-n, -n-2, -n-4$, etc. Away from these poles we can apply the formula (11.7.5), or generalizing it

$$(\nabla^2)^k \rho^\nu = \nu(\nu-2)\ldots(\nu-2k+2)(\nu+n-2)(\nu+n-4)\ldots \times (\nu+n-2k)\rho^{\nu-2k}. \tag{11.7.10}$$

Furthermore

$$\begin{aligned} \int_{R_n} \rho^\nu \phi(\mathbf{x})\, dX &= \frac{\Gamma(\tfrac{1}{2}\nu+1)\,\Gamma\{\tfrac{1}{2}(\nu+n)\}}{2^{2k}\Gamma(\tfrac{1}{2}\nu+k+1)\,\Gamma\{\tfrac{1}{2}(\nu+n)+k\}} \int_{R_n} \{(\nabla^2)^k \rho^{\nu+2k}\}\, \phi(\mathbf{x})\, dX \\ &= \frac{\Gamma(\tfrac{1}{2}\nu+1)\,\Gamma\{\tfrac{1}{2}(\nu+n)\}}{2^{2k}\Gamma(\tfrac{1}{2}\nu+k+1)\,\Gamma\{\tfrac{1}{2}(\nu+n)+k\}} \int_{R_n} \rho^{\nu+2k}\{(\nabla^2)^k \phi(\mathbf{x})\}\, dX. \end{aligned} \tag{11.7.11}$$

From (11.7.11), taking $k = 1$

$$\begin{aligned} \lim_{\nu+n\to 0} (\nu+n)\int_{R_n} \rho^\nu \phi(\mathbf{x})\, dX &= \frac{1}{2-n}\int_{R_n} (\nabla^2 \rho^{2-n})\, \phi(\mathbf{x})\, dX \\ &= +\Omega_n \phi(\mathbf{0}), \end{aligned} \tag{11.7.12}$$

and so the residue of ρ^ν at $-n$ is $+\Omega_n \delta_n$. Next, putting $k = 2$,

$$\begin{aligned} \lim_{\nu+n+2\to 0} (\nu+n+2)\int_{R_n} \rho^\nu \phi(\mathbf{x})\, dX &= \frac{1}{2n(2-n)}\int_{R_n} \{(\nabla^2)^2 \rho^{2-n}\}\, \phi(\mathbf{x})\, dX \\ &= \frac{1}{2n(2-n)}\int_{R_n} (\nabla^2 \rho^{2-n})(\nabla^2 \phi)\, dX \\ &= +\frac{\Omega_n}{2n}(\nabla^2 \phi)_{\mathbf{x}=0} \end{aligned} \tag{11.7.13}$$

and so we can define the residue of ρ^ν at $-n-2$ as $+(\Omega_n/2n)\nabla^2\delta_n$ and we can find the residues successively at each pole. In fact the residue of ρ^ν at $-n-2k$ appears as

$$+[2^k\{k!\}n(n+2)(n+2k-2)]^{-1}\Omega_n\{\nabla^2\}^k\delta_n. \tag{11.7.14}$$

11.8 The resolution of the Dirac function into plane waves

If $f \in L$, the space of one-dimensional locally integrable functions, if $\mathbf{x}$ is a vector in R_n and $\boldsymbol{\eta}$ a unit vector in R_n then $f(\boldsymbol{\eta}.\mathbf{x})$ is a locally integrable function in R_n. We say that $f(\boldsymbol{\eta}.\mathbf{x})$ is a 'plane wave' function.

We can find an orthogonal matrix A such that

$$A\boldsymbol{\eta} = \{1, 0, 0, \ldots, 0\} = \mathbf{i}.$$

Let also $A\mathbf{x} = \mathbf{y}$, so that, by the invariance under orthogonal transformation of scalar products,

$$\boldsymbol{\eta}.\mathbf{x} = \mathbf{i}.\mathbf{y} = y_1.$$

Thus if $\phi \in D_n$

$$\int_{R_n} f(\boldsymbol{\eta}.\mathbf{x})\,\phi(\mathbf{x})\,dX = \int_{-\infty}^{\infty} f(y_1)\,\Phi(\boldsymbol{\eta}, y_1)\,dy_1, \tag{11.8.1}$$

where

$$\Phi(\boldsymbol{\eta}, y_1) = \int_{\boldsymbol{\eta}.\mathbf{x}=y_1} \phi\{A^{-1}\mathbf{y}\}\,dy_2\,dy_3 \ldots dy_n,$$

and $\Phi(\boldsymbol{\eta}, y_1)$ is clearly of bounded support and belongs to C^∞, and so is a test function. The function Φ must also obviously depend on the vector $\boldsymbol{\eta}$. Thus is $f \in L$ we see that

$$\int_{R_n} f(\boldsymbol{\eta}.\mathbf{x})\,\phi(\mathbf{x})\,dX = \{f(t), \Phi(\boldsymbol{\eta}, t)\}. \tag{11.8.2}$$

We may generalize this to distributions. If F is a distribution in one dimension, we can define the plane wave distribution in n dimensions $F_{\boldsymbol{\eta}}$ to be

$$F\{\Phi\}. \tag{11.8.3}$$

Consider, for example, the one-dimensional distribution $|x|^\nu$. Now

$$\int_{-\infty}^{\infty} |x|^\nu\,\phi(x)\,dx = \int_0^\infty x^\nu\{\phi(x)+\phi(-x)\}dx.$$

$$= \int_0^\infty x^\nu\left\{\phi(x)+\phi(-x)-2H(1-x)\sum_{k=0}^{m-1}\frac{\phi^{2k}(0)x^{2k}}{\{(2k)!\}}\right\}dx$$

$$+2\sum_{k=0}^{m-1}\frac{\phi^{(2k)}(0)}{\{2k)!\}\{\nu+2k+1\}}. \tag{11.8.4}$$

Thus $|x|^\nu$ is a distribution, regular in ν except for poles at $\nu = -1$, -3, -5, etc., and the residue of $|x|^\nu$ at $\nu = -(2k+1)$ is $2\delta^{(2k)}/\{(2k)!\}$.

The corresponding plane wave distribution $|\boldsymbol{\eta}.\mathbf{x}|^\nu$ then also has simple poles at -1, -3, -5, etc.

Let us integrate $|\boldsymbol{\eta}.\mathbf{x}|^\nu$ over all unit vectors $\boldsymbol{\eta}$. This is an integration over the surface of the unit sphere in n dimensions, but to distinguish it from the sphere in the $\mathbf{x}$ space we shall denote it by ω_n.

Then

$$\int_{R_n}\left\{\int_{\omega_n} |\boldsymbol{\eta}.\mathbf{x}|^\nu \, d\omega_n\right\}\phi(\mathbf{x})\, dX = \int_{\omega_n}\left\{\int_{R_n} |\boldsymbol{\eta}.\mathbf{x}|^\nu \phi(\mathbf{x})\, dX\right\} d\omega_n. \tag{11.8.5}$$

Further, if we put $\boldsymbol{\eta}.\mathbf{x} = \rho \cos\theta$,

$$\begin{aligned}\int_{\omega_n} |\boldsymbol{\eta}.\mathbf{x}|^\nu \, d\omega_n &= \rho^\nu \int_0^\pi \Omega_{n-1} |\cos\theta|^\nu \sin\theta^{n-2}\, d\theta \\ &= \rho^\nu \Omega_{n-1}\int_0^1 t^{(\nu-1)/2}(1-t)^{(n-3)/2}\, dt \\ &= \rho^\nu \Omega_{n-1}\frac{\Gamma\{\frac{1}{2}(\nu+1)\}\,\Gamma\{\frac{1}{2}(n-1)\}}{\Gamma\{\frac{1}{2}(\nu+n)\}} \\ &= \frac{2\rho^\nu \pi^{(n-1)/2}\,\Gamma\{\frac{1}{2}(\nu+1)\}}{\Gamma\{\frac{1}{2}(\nu+n)\}}\end{aligned} \tag{11.8.6}$$

From (11.7.12) and (11.8.6)

$$\delta_n = \lim_{\nu+n\to 0}\frac{2\rho^\nu}{\Omega_n\,\Gamma\{\frac{1}{2}(\nu+n)\}} = \lim_{\nu+n\to 0}\tfrac{1}{2}(\pi)^{1/2-n}\frac{\Gamma(\frac{1}{2}n)}{\Gamma\{\frac{1}{2}(1+\nu)\}}\int_{\omega_n}|\boldsymbol{\eta}.\mathbf{x}|^\nu\, d\omega_n. \tag{11.8.7}$$

If n is even, say $n = 2m$, then there is no pole to be considered and

$$\delta_{2m} = (-1)^m\{(2m-1)!\}(2\pi)^{-2m}\int_{\omega_{2m}} |\boldsymbol{\eta}.\mathbf{x}|^{-2m}\, d\omega_{2m}, \tag{11.8.8}$$

and so the delta function in a space of $2m$ dimensions can be resolved into plane wave distributions.

When n is odd, $n = 2m+1$ say, there is a simple pole both of $|\boldsymbol{\eta}.\mathbf{x}|^\nu$ and of $\Gamma\{\frac{1}{2}(1+\nu)\}$ to be considered, and the ratio of their residues gives that

$$\delta_{2m+1} = (-1)^m\tfrac{1}{2}(2\pi)^{-2m}\int_{\omega_{2m+1}} \delta^{(2m)}\{|\boldsymbol{\eta}.\mathbf{x}|\}\, d\omega_{2m+1}. \tag{11.8.9}$$

The quantity in the integrand is the $2m$th distribution derivative of the one-dimensional delta distribution whose support is at $\boldsymbol{\eta}.\mathbf{x} = 0$. This resolution of the $(2m+1)$th dimensional delta function suggests Huyghens' principle.

11.9 The application to elliptic equations with constant coefficients

We shall, to end this chapter, apply some of these ideas to the solution of partial differential equations which are governed by a differential operator with constant coefficients, are of homogeneous degree and are elliptic. The operator therefore has the form

$$L(D), \Phi = C_{ij\ldots p}\frac{\partial^N \Phi}{\partial x_1{}^i\, \partial x_2{}^j \ldots \partial x_n{}^p}, \tag{11.9.1}$$

where $i, j \ldots, p$ run from 0 to N, $C_{ijk\ldots p}$ are constants which are zero unless $i + j + \ldots + p = N$, and the property of being elliptic implies that the algebraic form

$$L(\boldsymbol{\eta}) = C_{ij\ldots p}\eta_1{}^i\, \eta_2{}^j \ldots \eta_n{}^p$$

can only vanish if $\boldsymbol{\eta} = \mathbf{0}$. This, obviously, is only possible if N is even.

We shall say that a function $U(\mathbf{x})$ is a fundamental solution if

$$L(D).U(\mathbf{x}) = \delta_n. \tag{11.9.2}$$

From (11.8.7) we may put

$$U(\mathbf{x}) = \lim_{\nu+n\to 0} \tfrac{1}{2}\pi^{1/2-n}\Gamma\left(\frac{n}{2}\right)\int_{\omega_n} v(\boldsymbol{\eta}.\mathbf{x}, \nu)\, d\omega_n \tag{11.9.3}$$

where

$$L(D)v(\boldsymbol{\eta}.\mathbf{x}, \nu) = \frac{|\,\boldsymbol{\eta}.\mathbf{x}\,|^\nu}{\Gamma\{\frac{1}{2}(1+\nu)\}}. \tag{11.9.4}$$

Now

$$\frac{\partial}{\partial x_j}v(\boldsymbol{\eta}.\mathbf{x}, \nu) = \eta_j\left\{\frac{d}{dt}v(t, \nu)\right\}_{t=\boldsymbol{\eta}.\mathbf{x}}$$

and so from (11.9.4)

$$L(\boldsymbol{\eta})\frac{d^N}{dt^N}v(t, \nu) = \frac{|\,t\,|^\nu}{\Gamma\{\frac{1}{2}(1+\nu)\}}. \tag{11.9.5}$$

We wish to find a solution of this equation which is integrable over the sphere ω_n and is analytic in a domain of the ν-plane including $\nu = -n$. Clearly, since $L(\boldsymbol{\eta}) \neq 0$, there exist solutions of (11.9.5) which are analytic in the half-plane $\mathscr{R}(\nu) > 0$, and on ω_n these solutions will also be continuous functions of $\boldsymbol{\eta}$ and so integrable.

If we differentiate (11.9.5) twice with respect to t, we get

$$L(\boldsymbol{\eta})\frac{d^N}{dt^N}\left\{\frac{d^2 v(t, \nu)}{dt^2}\right\} = \frac{2\nu\,|\,t\,|^{\nu-2}}{\Gamma\{\frac{1}{2}(\nu-1)\}}. \tag{11.9.6}$$

Comparing (11.9.5) and (11.9.6) we see that if $\mathscr{R}(\nu) > 2$

$$2\nu v\{t, \nu - 2\} = \frac{d^2}{dt^2}v\{t, \nu\} \tag{11.9.7}$$

and (11.9.7) provides an analytic continuation from the solutions with $\mathscr{R}(\nu) > 0$ to all points of the ν plane excluding $\nu = 0, -2, -4$, etc.

If then n is odd, $n = 2m + 1$ say, there will be no difficulty. We shall have

$$2v\{t, -1\} = \frac{d^2}{dt^2}v\{t, 1\}. \tag{11.9.8}$$

Now from (11.9.5)

$$L(\eta)\frac{d^N}{dt^N}v\{t, 1\} = |t|, \tag{11.9.9}$$

and so since N is even

$$L(\eta)\, v\{t, 1\} = \frac{|t|^{N+1}}{\{(N+1)!\}}. \tag{11.9.10}$$

Since

$$\frac{d^2}{dt^2}|t| = 2\delta$$

it follows from (11.9.9) that

$$L(\eta)\frac{d^N}{dt^N}v\{t, -1\} = \delta, \tag{11.9.11}$$

and so $v\{t, -1\}$ is itself a fundamental solution of the ordinary differential equation (11.9.11). Then from (11.9.10):

$$\{(N-1)!\}\, L(\eta)\, v(t, -1) = |t|^{N-1}, \tag{11.9.12}$$

and from (11.9.7)

$$\{(2m)!\}v\{t, -2m-1\} = (-1)^m\{(m)!\}\frac{d^{2m}}{dt^{2m}}v\{t, -1\}. \tag{11.9.13}$$

There are therefore two cases to consider: if $N > 2m + 1$ then

$$\{(2m)!\}\{(N-2m-1)!\}\, L(\eta)\, v\{t, -2m-1\} = (-1)^m\{(m)!\}\, |t|^{N-2m-1}, \tag{11.9.14}$$

while if $N < (2m + 1)$

$$\{(2m)!\}\, L(\eta)\, v\{t, -2m-1\} = (-1)^m\{(m)!\}\delta^{2m-N}. \tag{11.9.15}$$

It is worth noticing that, of these two solutions, the first is locally integrable and the second is a distribution.

If n is even, $n = 2m$ say, we have to proceed with greater care. A purely formal particular integral of (11.9.5), in which we have put $N = 2M$ is

$$2^{2m}\Gamma\left\{M + \frac{1+\nu}{2}\right\}\left(\frac{\nu}{2} + M\right)\left(\frac{\nu}{2} + M - 1\right)\dots \times \left(\frac{\nu}{2} + 1\right) L(\eta)\, v\{t, \nu\} = |t|^{\nu+2M}, \tag{11.9.16}$$

and to this may be added a complementary function which will be a polynomial of degree not exceeding $2M - 1$.

If then $M \geqslant m$ we may put

$$2^{2m}\{(2M - 2m)!\}\{(m - 1)!\}\sqrt{\pi}\, L(\eta)\, v\{t, -2m\}$$
$$= (-1)^{m-1} \lim_{\nu+2m\to 0} \frac{|\, t\,|^{\nu+2M} - t^{2M-2m}}{\nu + 2m}$$
$$= (-1)^{m-1} t^{2(M-m)} \log \{|\, t\,|\}, \qquad (11.9.17)$$

which is locally integrable. To obtain the corresponding result when $M < m$, we have to introduce two new distributions. If we follow the notation of the theory of integral and meromorphic functions [20], we can define the locally integrable function $\log^+ x$, to be $\log x$ if $x > 0$, and zero if $x \leqslant 0$. This has distribution derivatives of all orders by the usual theorem, and so we may define the distribution $x_+{}^{-k}$, for positive integer k by

$$\frac{d^k}{dx^k} \log^+ x = (-1)^{k-1}\{(k - 1)!\} x_+{}^{-k}. \qquad (11.9.18)$$

We may then define the distribution $|\, x\,|^{-k}$ by the relation between functionals

$$|\, x\,|^{-k}\{\phi(x)\} = x_+{}^{-k}\{\phi(x) + \phi(-x)\}. \qquad (11.9.19)$$

Since this gives a formal definition to $|\, x\,|^{-k}$ for integer k, and since (11.9.16) is not singular as $\nu + 2m \to 0$, if $M < m$, we may substitute directly into (11.9.16). Hence if $M < m$

$$2^{2M}\Gamma\{M - m + \tfrac{1}{2}\}\, \Gamma\{m\}\, L(\eta)\, v\{t, -2m\}$$
$$= (-1)^M\, \Gamma\{(m - M)\}\, |\, t\,|^{-2(m-M)}. \qquad (11.9.20)$$

This solution is a distribution.

The substitution of (11.9.14), (11.9.15), (11.9.17) or (11.9.20) into (11.9.3) gives the fundamental solution $U(\mathbf{x})$, which therefore is locally integrable if $N \geqslant n$, and is a distribution if $N < n$.

If $\phi \in D_n$, the solution of

$$L(D)\Phi = \phi \qquad (11.9.21)$$

is

$$\Phi(\mathbf{x}_0) = \int_{R_n} U(\mathbf{x}-\mathbf{x}_0)\, \phi(\mathbf{x})\, dX. \qquad (11.9.22)$$

CHAPTER 12

Systems of Equations of Order greater than Two

12.1 Introduction

So far, we have concentrated almost solely on partial differential equations of the second order, or on pairs of simultaneous equations of first order. The theory of systems of order greater than two presents difficulties beyond the scope of this book, and the reader wishing to pursue the topic is recommended to read the relevant chapters in [1]. Since, however, some of the systems mentioned in Chapter 1 as being of immediate interest to the mathematical physicists are of order greater than two, we will now consider briefly how some of the concepts already encountered may be generalized. It is relevant to begin with a reminder that the only case where m first-order equations in n independent variables, for m unknown functions, can be considered equivalent to a single equation of order m for one function in n independent variables, is when $n = m = 2$.

12.2 Solution of the Cauchy problem

Let us begin by considering a system of m equations of first order in two independent variables, of the form

$$a_{ij}\frac{\partial \Phi_j}{\partial x} + b_{ij}\frac{\partial \Phi_j}{\partial y} + c_i = 0, i = 1, 2, \ldots, m. \qquad (12.2.1)$$

These equations are linear if the coefficients a_{ij}, b_{ij} are functions of x and y only, and c_i is a linear function of Φ_j; they are semi-linear if a_{ij} and b_{ij} depend on x and y only and c_i depends on x, y and Φ_j, the dependence on Φ_j being of a non-linear character, and they are quasi-linear if a_{ij}, b_{ij} and c_i are functions of x, y and Φ_j.

Let

$$f(x, y) = 0 \qquad (12.2.2)$$

be the implicit equation of a curve C at each point of which $\partial f/\partial x$ and $\partial f/\partial y$ exist. If a function Φ_j is known on C, then its tangential derivative is known along C, and if we call this u_j, then

$$\frac{\partial f}{\partial y}\frac{\partial \Phi_j}{\partial x} - \frac{\partial f}{\partial x}\frac{\partial \Phi_j}{\partial y} + \frac{\partial f}{\partial y}u_j = 0 \tag{12.2.3}$$

along C. If we eliminate $\partial \Phi_j/\partial y$ between (12.2.1) and (12.2.3) we find that

$$\frac{\partial \Phi_j}{\partial x}\left\{a_{ij}\frac{\partial f}{\partial x} + b_{ij}\frac{\partial f}{\partial y}\right\} + b_{ij}\frac{\partial f}{\partial y}u_j + c_i\frac{\partial f}{\partial x} = 0. \tag{12.2.4}$$

Hence if the determinant

$$\left| a_{ij}\frac{\partial f}{\partial x} + b_{ij}\frac{\partial f}{\partial y} \right| \neq 0,$$

(12.2.4) can be solved for the partial derivatives $\partial \Phi_j/\partial x$, and when these are known the other derivatives $\partial \Phi_j/\partial y$ may be found from (12.2.3).

Alternatively

$$\left| a_{ij}\frac{\partial f}{\partial x} + b_{ij}\frac{\partial f}{\partial y} \right| = 0 \tag{12.2.5}$$

is a partial differential equation of first order and degree m for families of curves $f(x, y) = \text{const}$. Any curve which satisfies this differential equation we call a characteristic. Any curve which has partial differentials $\partial f/\partial x$ and $\partial f/\partial y$ at each point, and does not satisfy (12.2.5), we say is a free curve.

We then see that if, for a given curve C, we can solve the Cauchy problem (which is that if Φ is known on C, we can find Φ in the neighbourhood of C), C must be free.

Corresponding to any real root λ of the equation

$$| a_{ij} - \lambda b_{ij} | = 0 \tag{12.2.6}$$

will be the real characteristic with equation

$$\frac{\partial f}{\partial y} + \lambda\frac{\partial f}{\partial x} = 0. \tag{12.2.7}$$

The non-real characteristics will go in pairs of complex conjugate roots corresponding to (12.2.6). Hence it is clear that any system with an odd order of equations must have at least one real characteristic.

If (12.2.6) has no real roots, we say that the system (12.2.1) is elliptic, while if (12.2.6) has m real distinct characteristics we say that it is totally hyperbolic. In practice, it is not always important that all the real roots should be distinct, provided that there are at least two distinct real characteristics.

On each real characteristic, by substituting from (12.2.7) in (12.2.4), we get a total differential relation between the Φ_j on the characteristics. If therefore the system is totally hyperbolic, and the values of Φ_j are known on a free curve C, the system can formally be integrated. If the system is elliptic then m is even, and it seems that the correct boundary data will consist of $m/2$ pieces of independent information at every point of a closed boundary. If the system is neither elliptic nor totally hyperbolic the question of the required data for a correctly posed Cauchy problem is much more complex.

We can introduce similar definitions for a single equation of order m

$$\sum_{k=0}^{m} \alpha_k \frac{\partial^m \Phi}{\partial x^k\, \partial y^{m-k}} + \beta = 0. \tag{12.2.8}$$

For this to be linear or semi-linear, the coefficients α_k must depend on x and y only, while for linearity β depends on x and y and linearly on Φ and its partial derivatives up to order $m - 1$, while if the dependence is non-linear we have a semi-linear equation. If the α_k and β depend non-linearly on Φ and its partial derivatives of order up to $(m - 1)$ and on x and y, then we have a quasi-linear equation. The equation corresponding to (12.2.5) is

$$\sum_{k=0}^{m} \alpha_k \left(\frac{\partial f}{\partial x}\right)^k \left(\frac{\partial f}{\partial y}\right)^{m-k} = 0 \tag{12.2.9}$$

and the other definitions follow directly.

12.3 Generalization to *n* independent variables

We can also extend these ideas to m first-order equations in n independent variables. Consider the system

$$a_{ijk}\frac{\partial \Phi_j}{\partial x_k} + c_i = 0, \tag{12.3.1}$$

where the distinction between linear, semi-linear and quasi-linear is obvious.

Now take a surface with the equation

$$f(\mathbf{x}) = 0. \tag{12.3.2}$$

We can take a section of this surface by the plane

$$x_2 = \text{const},\ x_3 = \text{const.}, \ldots, x_{n-1} = \text{const.}$$

The section will be a plane curve, and on it we shall have

$$\frac{\partial \Phi_j}{\partial x_1}\frac{\partial f}{\partial x_n} - \frac{\partial \Phi_j}{\partial x_n}\frac{\partial f}{\partial x_1} = u_{j1} \tag{12.3.3}$$

if Φ_j is known on the surface. We may similarly derive

$$\frac{\partial \Phi_j}{\partial x_k}\frac{\partial f}{\partial x_n} - \frac{\partial \Phi_j}{\partial x_n}\frac{\partial f}{\partial x_k} = u_{jk} \tag{12.3.4}$$

for $k = 2, \ldots, n - 1$. We may use these to eliminate the partial derivatives

$$\frac{\partial \Phi_j}{\partial x_k}$$

$j = 1, \ldots, m$, $k = 1, \ldots, n - 1$ from (12.3.1) and we find that, if we put $u_{jn} = 0$,

$$\frac{\partial \Phi_j}{\partial x_n}\left\{a_{ijk}\frac{\partial f}{\partial x_k}\right\} + c_i\frac{\partial f}{\partial x_n} + a_{ijk}u_{jk} = 0. \tag{12.3.5}$$

This is soluble for $\partial\Phi_j/\partial x_n$ if and only if

$$\left| a_{ijk}\frac{\partial f}{\partial x_k} \right| \neq 0.$$

A surface satisfying this condition will therefore be required for the boundary data for Cauchy's problem.

A family of surfaces $f(\mathbf{x}) =$ const. which satisfy

$$\left| a_{ijk}\frac{\partial f}{\partial x_k} \right| = 0 \tag{12.3.6}$$

at every point and for each value of the constant, are characteristic surfaces. Alternatively (12.3.6) will have a system of integral curves, which are called its characteristic rays, or bicharacteristics. Through any given point there will, in general, pass a sheaf of such rays, and they will lie in a hypersurface which we may call, by analogy, the characteristic conoid of the point. A characteristic surface is one which contains a bicharacteristic through each of its points, and so touches the characteristic conoid along this ray.

An equation will be elliptic if it has no real bicharacteristics and this, again, implies that m must be even. It is totally hyperbolic if (12.3.6) possesses m distinct roots for $\partial f/\partial x_n$, for arbitrarily given values of $\partial f/\partial x_1$, $\partial f/\partial x_2, \ldots, \partial f/\partial x_{n-1}$, although it is frequently unimportant if some of the roots do actually coincide.

12.4 The adjoint equation system

We may finally define an adjoint system to the set of linear equations above. We consider the system of differential expressions as components of a vector

$$L\{\mathbf{\Phi}\} = \left\{a_{ijk}\frac{\partial \Phi_j}{\partial x_k} + b_{ij}\Phi_j\right\} = -c_i, \tag{12.4.1}$$

where now the a_{ijk}, b_{ij} and c_i are functions of the vector $\mathbf{x}$ only. Then we define the adjoint equation

$$M\{\boldsymbol{\Psi}\} = 0 \tag{12.4.2}$$

by the requirement that

$$\boldsymbol{\Psi}.L\{\boldsymbol{\Phi}\} - \boldsymbol{\Phi}.M\{\boldsymbol{\Psi}\} = \operatorname{div} \boldsymbol{\Theta}, \tag{12.4.3}$$

where $\boldsymbol{\Theta}$ is a vector and is a bilinear form in the partial derivatives of $\boldsymbol{\Phi}$ and $\boldsymbol{\Psi}$.

It is clear that

$$M\{\boldsymbol{\Psi}\} \equiv \left\{-\frac{\partial}{\partial x_k}(a_{ijk}\boldsymbol{\Psi}_j) + b_{ij}\boldsymbol{\Psi}_j\right\} = 0, \tag{12.4.4}$$

as then

$$\boldsymbol{\Psi}.L\{\boldsymbol{\Phi}\} - \boldsymbol{\Phi}.M\{\boldsymbol{\Psi}\} = \frac{\partial}{\partial x_k}\{a_{ijk}\Psi_i\Phi_j\}. \tag{12.4.5}$$

A system of equations which is its own adjoint is said to be self-adjoint. The condition that the system should be self-adjoint is that a_{ijk} should be symmetric in i and j and that

$$\frac{\partial a_{ijk}}{\partial x_k} = 2b_{ij}. \tag{12.4.6}$$

CHAPTER 13

The Equations of Fluid Dynamics

13.1 The linear cone field

We shall finally illustrate some of the theory of partial differential equations, by applying it to the equations of fluid dynamics which were stated in Section 1.5. It was pointed out in that section that the equations of the irrotational motion of an inviscid, incompressible fluid were reducible to the Laplace equation, while the approximate equations for a small perturbation of an inviscid, compressible fluid were reducible either to the Laplace or to the wave equation. Since we have dealt at length with these equations in earlier chapters we shall not discuss them further. There is, however, just one matter in this connection which is worthy of discussion.

It was shown that the perturbation potential for a uniform stream of velocity U_0 in the x direction of an inviscid compressible fluid was

$$(1 - M_0{}^2)\frac{\partial^2\phi}{\partial x^2} + \frac{\partial^2\phi}{\partial y^2} + \frac{\partial^2\phi}{\partial z^2} = 0, \qquad (13.1.1)$$

where M_0 was the 'Mach' number of the uniform stream, that is $a_0M_0 = U_0$, where a_0 was the velocity of sound in the undisturbed stream.

Suppose that the body producing the disturbance is a cone of small cross-section, whose generators all meet at a vertex which we may, without loss of generality, assume to be the origin. The quantities characterizing the inviscid problem are the cone angle, the velocity U_0, and the pressure, density and temperature of the undisturbed stream. No combination of these has the dimension of length, and so no solution can depend on a length. Thus all physical quantities must be constant on rays through the vertex of the cone. Now if $M_0 < 1$, (13.1.1) is reducible, by an affine transformation, to the Laplace equation, and we have seen that the effects of the boundary conditions for this equation are felt at all points of space. Thus, the disturbance created by the cone cannot decay at infinity, since it is constant along these rays, and this violates the boundary condition at infinity. There can therefore be no solution of the steady inviscid perturbation problem with $M_0 < 1$.

Physically, in subsonic flow, a separation bubble will occur at the vertex in steady flow. This will be small, and so the disturbances to the flow will still be small, but it will be of finite and definite curvature. Thus a scale of length is required, and this can only be introduced by considering the viscosity.

If $M_0 > 1$, put $\beta^2 = M_0{}^2 - 1$. Since our solution does not depend on length, we may introduce dimensionless independent variables

$$\xi = \frac{\beta y}{x} \quad \text{and} \quad \eta = \frac{\beta z}{x}. \tag{13.1.2}$$

In fact we are not justified in asserting that $\phi(x, y, z) = \phi(\xi, \eta)$, as the velocity potential is not a true physical variable. If, however, we consider, say, $u = -\partial\phi/\partial x$, then u is the disturbance caused to the velocity component in the x-direction, and is a true physical variable. It also is clear that u will satisfy the same equation if (13.1.1) is differentiated partially with respect to x.

The substitution of the new variables (13.1.2) gives as an alternative equation for u:

$$(\xi^2 - 1)\frac{\partial^2 u}{\partial \xi^2} + 2\xi\eta\frac{\partial^2 u}{\partial \xi\, \partial \eta} + (\eta^2 - 1)\frac{\partial^2 u}{\partial \eta^2} + 2\xi\frac{\partial u}{\partial \xi} + 2\eta\frac{\partial u}{\partial \eta} = 0, \tag{13.1.3}$$

which is an equation we considered for illustrative purposes in Chapter 3. It would be out of place here to consider the actual solution of physical problems, but the reader is referred to [10] or [21] where the solution for the flow about a thin delta-shaped aerofoil is found with considerable mathematical elegance.

There is just one further point which is worth stressing. With $M_0 > 1$, (13.1.1) is the wave equation, and so is certainly of normal hyperbolic type. On the other hand, from Chapter 3 we see that (13.1.3) is always elliptic when $\xi^2 + \eta^2 < 1$, that is in physical terms, within the Mach cone.

13.2 Inviscid compressible steady flow in two dimensions

Let us now consider the equations for the steady motion in two dimensions of an inviscid compressible fluid. From Section 1.5 we may take these as

$$u\frac{\partial u}{\partial x} + v\frac{\partial u}{\partial y} + \frac{1}{\rho}\frac{\partial p}{\partial x} = 0, \qquad (13.2.1)$$

$$u\frac{\partial v}{\partial x} + v\frac{\partial v}{\partial y} + \frac{1}{\rho}\frac{\partial p}{\partial y} = 0, \qquad (13.2.2)$$

$$\frac{\partial u}{\partial x} + \frac{\partial v}{\partial y} + \frac{u}{\rho}\frac{\partial \rho}{\partial x} + \frac{v}{\rho}\frac{\partial \rho}{\partial y} = 0, \qquad (13.2.3)$$

$$u\frac{\partial S}{\partial x} + v\frac{\partial S}{\partial y} = 0, \qquad (13.2.4)$$

together with an equation of state, $p = p(\rho, S)$. This system is therefore quasi-linear.

If we put $\zeta = dy/dx$, the determinantal equation for the characteristics is

$$\begin{vmatrix} v - \zeta u & 0 & -\zeta & 0 \\ 0 & v - \zeta u & 1 & 0 \\ -a^2\zeta & a^2 & v - \zeta u & 0 \\ -b^2\zeta & b^2 & 0 & v - \zeta u \end{vmatrix} = 0 \qquad (13.2.5)$$

where we have put

$$a^2 = \partial p/\partial \rho,\ b^2 = (1/\rho)\partial p/\partial S \quad \text{and} \quad \partial p/\partial x = a^2 \partial \rho/\partial x + \rho b^2 \partial S/\partial x.$$

This simplifies to

$$(v - \zeta u)^2\{(v - \zeta u)^2 - a^2(1 + \zeta^2)\} = 0. \qquad (13.2.6)$$

There are therefore always at least two real, coincident characteristics corresponding to the root $v - \zeta u = 0$. Physically these are the stream lines. On each of these two, although they coincide, there should be a differential relation between the dependent variables. The first of these relations is given immediately by (13.2.4), that is on the stream lines

$$dS = 0. \qquad (13.2.7)$$

The second relation may also be obtained from the equations. If we multiply (13.2.1) by u, (13.2.2) by v and add, we obtain

$$\tfrac{1}{2}d(u^2 + v^2) + \frac{dp}{\rho} = 0, \qquad (13.2.8)$$

on the stream line.

The remaining quadratic is

$$\zeta^2(u^2 - a^2) - 2\zeta uv + v^2 - a^2 = 0,$$

which has real roots if $u^2 + v^2 > a^2$ (the velocity is supersonic), and conjugate complex roots if $u^2 + v^2 < a^2$ (the velocity is subsonic). When the characteristics are real they are seen to be lines inclined to the stream lines at angles $\pm\mu$, where $(u^2 + v^2)\sin^2\mu = a^2$. These are the so-called 'Mach' lines, and μ is the Mach angle. If we put $u = q \cos \theta$, $v = q \sin \theta$, so that the two characteristics are given by the equations

$$\frac{dy}{dx} = \tan(\theta \pm \mu) \tag{13.2.9}$$

we may see that the differential relations on these lines may be obtained by applying to the four equations (13.2.1)–(13.2.4) the multiplying factors $-\sin(\theta \pm \mu)$, $\cos(\theta \pm \mu)$, $\pm q \sin \mu$ and $\pm b^2/q \sin \mu$ respectively. These differential relations are then

$$\rho q^2 d\theta \pm \cot \mu \, dp = 0. \tag{13.2.10}$$

For irrotational flows with constant entropy S, the last equation (13.2.4) becomes redundant, and $dp/\rho q^2$ can be expressed in terms of μ only, so that (13.2.10) is integrable. With a suitable choice of constant of integration, it can be replaced by

$$d(\theta \mp \omega) = 0, \tag{13.2.11}$$

where

$$d\omega = -\sin \mu \cos \mu \frac{d\rho}{\rho}. \tag{13.2.12}$$

In the particular case of an adiabatic gas wherein

$$p \propto \rho^\gamma$$

then

$$\omega = \mu - \frac{\pi}{2} + \left(\frac{\gamma+1}{\gamma-1}\right)^{1/2} \tan^{-1}\left\{\left(\frac{\gamma-1}{\gamma+1}\right)^{1/2} \cot \mu\right\}, \tag{13.2.13}$$

and is called the Prandtl-Meyer angle.

We have seen previously that for these irrotational flows we may introduce a velocity potential function $\phi(x, y)$ with the property that $u = -\partial\phi/\partial x$ and $v = -\partial\phi/\partial y$. The equation for this potential function is

$$(a^2 - u^2)\frac{\partial^2\phi}{\partial x^2} - 2uv\frac{\partial^2\phi}{\partial x\, \partial y} + (a^2 - v^2)\frac{\partial^2\phi}{\partial y^2} = 0. \tag{13.2.14}$$

This will be elliptic or hyperbolic according as $u^2 + v^2$ is less or greater than a^2, and in the latter case it will have the characteristics (13.2.9), with the differential conditions (13.2.11).

We may further introduce the Legendre potential function $\Phi(u, v)$ defined by

$$\Phi(u, v) = ux + vy + \phi(x, y). \tag{13.2.15}$$

This has the properties that

$$\frac{\partial\Phi}{\partial u} = x + u\frac{\partial x}{\partial u} + v\frac{\partial y}{\partial u} + \frac{\partial\phi}{\partial x}\frac{\partial x}{\partial u} + \frac{\partial\phi}{\partial y}\frac{\partial y}{\partial u} = x, \tag{13.2.16}$$

and similarly

$$\frac{\partial \Phi}{\partial v} = y. \tag{13.2.17}$$

Since also

$$1 = \frac{\partial x}{\partial u}\frac{\partial u}{\partial x} + \frac{\partial x}{\partial v}\frac{\partial v}{\partial x}$$

and

$$0 = \frac{\partial x}{\partial u}\frac{\partial u}{\partial y} + \frac{\partial x}{\partial v}\frac{\partial v}{\partial y},$$

then

$$\frac{\partial^2 \Phi}{\partial u^2} = \frac{\partial x}{\partial u} = \mathcal{J}^{-1}\frac{\partial v}{\partial y} = -\mathcal{J}^{-1}\frac{\partial^2 \phi}{\partial y^2},$$

and

$$\frac{\partial^2 \Phi}{\partial u\,\partial v} = \mathcal{J}^{-1}\frac{\partial^2 \phi}{\partial x\,\partial y}$$

where $\mathcal{J}$ is the Jacobian

$$\frac{\partial u}{\partial x}\frac{\partial v}{\partial y} - \frac{\partial u}{\partial y}\frac{\partial v}{\partial x}.$$

Provided then that $\mathcal{J}$ does not vanish, $\Phi(u, v)$ will satisfy the linear second-order equations:

$$(a^2 - v^2)\frac{\partial^2 \Phi}{\partial u^2} + 2uv\frac{\partial^2 \Phi}{\partial u\,\partial v} + (a^2 - u^2)\frac{\partial^2 \Phi}{\partial v^2} = 0. \tag{13.2.18}$$

This will be elliptic or hyperbolic according as $u^2 + v^2$ is less than, or greater than a^2, and in the latter case its characteristics will be given by (13.2.11) and the differential relation satisfied on them will be (13.2.9). It is shown, in [2] for example, that in the case of an adiabatic gas these characteristics are epicycloids, and are orthogonal to the corresponding characteristic of (13.2.14).

In fact the benefit to be derived from having a linear equation (13.2.18) rather than a quasi-linear equation to solve is much less than has been hoped at various times. The mapping of the physical plane, in which x and y are coordinates, on to the hodograph plane, in which u and v are the coordinates, is sufficiently complicated to offset any advantage gained from the simpler equation. The reader wishing to pursue this point is recommended to the chapter on the hodograph equation in [10].

In the case of an adiabatic gas $p \propto \rho^\gamma$, we may supplement (13.2.18) with the so-called 'Bernoulli' equation

$$2a^2 + (\gamma - 1)(u^2 + v^2) = 2h, \tag{13.2.19}$$

where h is a constant. If we assume that a and u are approximately equal (transonic flow) we may put approximately

$$u = a_*(1+\xi),\ v = a_*\eta,\ a = a_*\left(1 - \frac{\gamma-1}{2}\xi\right)$$

where if ξ is small it is assumed that η is $o(\xi)$. Then (13.2.18) can be approximated by the Tricomi equation

$$\frac{\partial^2\Phi}{\partial u^2} = (\gamma+1)u\frac{\partial^2\Phi}{\partial v^2}, \tag{13.2.20}$$

which is effectively the equation we considered as an example in Chapter 4.

13.3 Inviscid compressible steady flow of an electrically conducting fluid

Let us still consider two-dimensional steady flows of a compressible inviscid fluid at constant entropy, but now assume that the fluid is electrically conducting. Equation (13.2.3) remains as before, but (13.2.1) and (13.2.2) are replaced by:

$$u\frac{\partial u}{\partial x} + v\frac{\partial u}{\partial y} + \frac{1}{\rho}\frac{\partial p}{\partial x} + \frac{\mu_e}{\rho}H_y\left\{\frac{\partial H_y}{\partial x} - \frac{\partial H_x}{\partial y}\right\} = 0, \tag{13.3.1}$$

and

$$u\frac{\partial v}{\partial x} + v\frac{\partial v}{\partial y} + \frac{1}{\rho}\frac{\partial p}{\partial y} - \frac{\mu_e}{\rho}H_x\left\{\frac{\partial H_y}{\partial x} - \frac{\partial H_x}{\partial y}\right\} = 0 \tag{13.3.2}$$

and in addition we have the two equations

$$\frac{\partial H_x}{\partial x} + \frac{\partial H_y}{\partial y} = 0, \tag{13.3.3}$$

and

$$uH_y - vH_x - \nu_H\left\{\frac{\partial H_y}{\partial x} - \frac{\partial H_x}{\partial y}\right\} = \text{const.} \tag{13.3.4}$$

These have interesting properties, for if $\nu_H \neq 0$, we can use (13.3.4) to eliminate from (13.3.1) and (13.3.2) the partial derivatives of H_x and H_y, and the resulting equations taken with (13.2.1) will then have the same classification as the ordinary equations of fluid dynamics, since this is determined by the highest-order partial derivatives. In addition, taking together (13.3.3) and (13.3.4) we see that these together form an elliptic system. In fact if $H_x = \partial\psi/\partial y$, $H_y = -\partial\psi/\partial x$, which satisfies (13.3.3), then

$$\nu_H\nabla^2\psi - u\frac{\partial\psi}{\partial x} - v\frac{\partial\psi}{\partial y} = 0. \tag{13.3.5}$$

If the fluid is of infinite conductivity however, which is an approximation frequently made, then $\nu_H = 0$, and the behaviour of the system is changed. From (13.3.4) we may then put $H_x = uh$, $H_y = vh$, and from (13.3.3)

$$\left(\frac{\partial u}{\partial x} + \frac{\partial v}{\partial y}\right) + \frac{u}{h}\frac{\partial h}{\partial x} + \frac{v}{h}\frac{\partial h}{\partial y} = 0. \qquad (13.3.6)$$

If this is subtracted from (13.2.3) we see that

$$u\frac{\partial}{\partial x}\log\left(\frac{\rho}{h}\right) + v\frac{\partial}{\partial y}\log\left(\frac{\rho}{h}\right) = 0 \qquad (13.3.7)$$

and so the stream line appears as an extra real characteristic on which ρ/h is constant.

This is therefore yet another example of the phenomenon which is not uncommon in mathematical physics, where, although a parameter may in practice be very small, yet the approximation of putting it equal to zero changes the character of the problem.

13.4 Inviscid compressible steady flow in three dimensions

If we look at the steady three-dimensional motion of an inviscid compressible fluid at constant entropy, we may again introduce the velocity potential, which will now satisfy the partial differential equation

$$(a^2 - u^2)\frac{\partial^2\phi}{\partial x^2} + (a^2 - v^2)\frac{\partial^2\phi}{\partial y^2} + (a^2 - w^2)\frac{\partial^2\phi}{\partial z^2} - 2vw\frac{\partial^2\phi}{\partial y\,\partial z}$$
$$- 2wu\frac{\partial^2\phi}{\partial z\,\partial x} - 2uv\frac{\partial^2\phi}{\partial x\,\partial y} = 0 \qquad (13.4.1)$$

wherein

$$u = -\frac{\partial\phi}{\partial x}, \quad v = -\frac{\partial\phi}{\partial y}, \quad \text{and} \quad w = -\frac{\partial\phi}{\partial z}.$$

To determine the properties of this system we must look therefore at the quadratic form:

$$(a^2 - u^2)\xi_1^2 + (a^2 - v^2)\xi_2^2 + (a^2 - w^2)\xi_3^2$$
$$- 2vw\xi_2\xi_3 - 2wu\xi_3\xi_1 - 2uv\xi_1\xi_2 = 0.$$

Sufficient conditions for this to be positive definite are (see for example [3])

$$a^2 > u^2,$$
$$a^2 > u^2 + v^2$$

and

$$a^2 > u^2 + v^2 + w^2.$$

Clearly the last condition is sufficient to ensure that the other two are satisfied. The condition that (13.4.1) is elliptic is therefore that

$a^2 > u^2 + v^2 + w^2$, or in other words that the velocity is subsonic. The coefficient matrix is reducible by orthogonal transformation to the normal form:

$$\begin{pmatrix} a^2 & 0 & 0 \\ 0 & a^2 & 0 \\ 0 & 0 & a^2 - u^2 - v^2 - w^2 \end{pmatrix}$$

and hence if the velocity is supersonic, that is $u^2 + v^2 + w^2 > a^2$, the equation (13.4.1) is of normal hyperbolic type.

It is fairly clear that in this case the characteristic cone is a right circular cone of semi-angle μ where μ the Mach angle is defined by $(u^2 + v^2 + w^2) \sin^2 \mu = a^2$, and the axis of the cone is a tangent to the stream direction

$$\frac{dx}{u} = \frac{dy}{v} = \frac{dz}{w}.$$

The bicharacteristics are twisted curves whose tangents are everywhere generators of the characteristic cone at the point. The set of bicharacteristics through each point generates the characteristic conoid.

13.5 Unsteady one-dimensional flow

Let us now consider the equations of the unsteady one-dimensional motion of a compressible inviscid gas. These are:

$$\frac{\partial \rho}{\partial t} + u\frac{\partial \rho}{\partial x} + \rho\frac{\partial u}{\partial x} = 0, \qquad (13.5.1)$$

and

$$\frac{\partial u}{\partial t} + u\frac{\partial u}{\partial x} + \frac{1}{\rho}\frac{\partial p}{\partial x} = 0. \qquad (13.5.2)$$

If we once more replace $\partial p/\partial x$ by $a^2 \partial \rho/\partial x$, we see that these may be combined to give

$$\frac{\partial u}{\partial t} + (u \pm a)\frac{\partial u}{\partial x} \pm \frac{a}{\rho}\left\{\frac{\partial \rho}{\partial t} + (u \pm a)\frac{\partial \rho}{\partial x}\right\} = 0.$$

The characteristics are therefore always real and have the equations

$$\frac{dx}{dt} = u \pm a. \qquad (13.5.3)$$

The corresponding differential relations which are to be satisfied on them are:

$$du \pm \frac{a\, d\rho}{\rho} = 0. \qquad (13.5.4)$$

Since we are considering flows at constant entropy, the equations (13.5.4) can be integrated to give

$$u \pm A(\rho) = \text{const.} \tag{13.5.5}$$

If the flow is adiabatic so that $p \propto \rho^\gamma$ then

$$A(\rho) = \frac{2}{\gamma - 1} a. \tag{13.5.6}$$

If we introduce the 'Riemann invariants'

$$2r = A(\rho) + u \quad \text{and} \quad 2s = A(\rho) - u,$$

we obviously may assert that r is constant on each characteristic $dx/dt = u + a$, while s is constant on each characteristic $dx/dt = u - a$. It follows then that on $dx/dt = (u + a)$ only s varies, and so we may express this alternatively in the form

$$\frac{\partial x}{\partial s} = (u + a)\frac{\partial t}{\partial s}, \tag{13.5.7}$$

and by an exactly similar argument

$$\frac{\partial x}{\partial r} = (u - a)\frac{\partial t}{\partial r}, \tag{13.5.8}$$

so that we have replaced the original equations by a pair of linear, first-order, simultaneous partial differential equations for x and t. From (13.5.7) we derive

$$\frac{\partial^2 x}{\partial r\,\partial s} = (u + a)\frac{\partial^2 t}{\partial r\,\partial s} + \left(\frac{\partial u}{\partial r} + \frac{\partial a}{\partial r}\right)\frac{\partial t}{\partial s}, \tag{13.5.9}$$

and from (13.5.8)

$$\frac{\partial^2 x}{\partial s\,\partial r} = (u - a)\frac{\partial^2 t}{\partial s\,\partial r} + \left(\frac{\partial u}{\partial s} - \frac{\partial a}{\partial s}\right)\frac{\partial t}{\partial r}. \tag{13.5.10}$$

If we then make the usual assumption that the two mixed second partial derivatives are equal, we may obtain a single equation of second order for t, namely

$$2a\frac{\partial^2 t}{\partial r\,\partial s} + \left(\frac{\partial u}{\partial r} + \frac{\partial a}{\partial r}\right)\frac{\partial t}{\partial s} + \left(\frac{\partial a}{\partial s} - \frac{\partial u}{\partial s}\right)\frac{\partial t}{\partial r} = 0. \tag{13.5.11}$$

We notice that

$$u = r - s$$

while a, the local velocity of sound, is a function of $r + s$. If we denote by a' the derivative of a with respect to $r + s$, then equation (13.5.11) takes the simplified form:

$$2a\frac{\partial^2 t}{\partial r\,\partial s} + (1 + a')\left(\frac{\partial t}{\partial r} + \frac{\partial t}{\partial s}\right) = 0. \tag{13.5.12}$$

In the particular case of an adiabatic gas, (13.5.12) takes the even simpler form

$$2(\gamma - 1)(r + s)\frac{\partial^2 t}{\partial r\, \partial s} + (\gamma + 1)\left(\frac{\partial t}{\partial r} + \frac{\partial t}{\partial s}\right) = 0. \quad (13.5.13)$$

This is the Darboux equation which we examined as an example in Chapter 10. It is further of interest to notice that simple considerations of kinetic theory lead to the conclusion that, for a monatomic gas, γ ought to be 5/3, and for a diatomic gas, 7/5. The value $\gamma = 5/3$ gives $(\gamma + 1)/2(\gamma - 1) = 2$, while $\gamma = 7/5$ leads to $(\gamma + 1)/2(\gamma - 1) = 3$. We are led therefore to examine further the Darboux equation

$$(r + s)\frac{\partial^2 t}{\partial r\, \partial s} + N\left\{\frac{\partial t}{\partial r} + \frac{\partial t}{\partial s}\right\} = 0, \quad (13.5.14)$$

where N is an integer. Obviously when $N = 0$ the equation is quite simply the wave equation, while when $N = 1$ it can be written

$$\frac{\partial^2}{\partial r\, \partial s}\{(r + s)t\} = 0, \quad (13.5.15)$$

and so if we adopt D'Alembert's solution of the wave equation, the solution of (13.5.15) is:

$$t(r, s) = (r + s)^{-1}\{f(r) + g(s)\}, \quad (13.5.16)$$

where f and g are arbitrary functions. Now let us differentiate (13.5.14) with respect to both r and s. As a consequence of (13.5.15) we obtain

$$(r + s)\frac{\partial^4 t}{\partial r^2\, \partial s^2} + (N + 1)\left(\frac{\partial^3 t}{\partial r^2\, \partial s} + \frac{\partial^3 t}{\partial r\, \partial s^2}\right) = 0. \quad (13.5.17)$$

Hence if $t_N(r, s)$ is a solution of (13.5.14) with parameter N, $\partial^2 t_N/\partial r\, \partial s$ is a solution of the same equation with the parameter raised to $(N + 1)$. Hence by induction

$$t_N(r, s) = \frac{\partial^{N-1}}{\partial r^{N-1}}\{(r + s)^{-N} f(r)\} + \frac{\partial^{N-1}}{\partial s^{N-1}}\{(r + s)^{-N} g(s)\}, \quad (13.5.18)$$

where again f and g are arbitrary, but not the same, functions.

Taking into account the effects of a magnetic field on a conducting one-dimensional compressible fluid adds nothing new in principle to what we have just done, but the reader interested in further details is referred to [13].

13.6 Unsteady three-dimensional flow

We shall in conclusion look at the equations for the unsteady three-dimensional motion of a compressible inviscid fluid at constant entropy.

The equations are

$$\frac{\partial\rho}{\partial t}+u\frac{\partial\rho}{\partial x}+v\frac{\partial\rho}{\partial y}+w\frac{\partial\rho}{\partial z}+\rho\left\{\frac{\partial u}{\partial x}+\frac{\partial v}{\partial y}+\frac{\partial w}{\partial z}\right\}=0, \quad (13.6.1)$$

$$\frac{\partial u}{\partial t}+u\frac{\partial u}{\partial x}+v\frac{\partial u}{\partial y}+w\frac{\partial u}{\partial z}+\frac{a^2}{\rho}\frac{\partial p}{\partial x}=0, \quad (13.6.2)$$

$$\frac{\partial v}{\partial t}+u\frac{\partial v}{\partial x}+v\frac{\partial v}{\partial y}+w\frac{\partial v}{\partial z}+\frac{a^2}{\rho}\frac{\partial p}{\partial y}=0 \quad (13.6.3)$$

and

$$\frac{\partial w}{\partial t}+u\frac{\partial w}{\partial x}+v\frac{\partial w}{\partial y}+w\frac{\partial w}{\partial z}+\frac{a^2}{\rho}\frac{\partial p}{\partial z}=0. \quad (13.6.4)$$

From the previous chapter we see therefore that the equations of the characteristic surfaces, if they exist, are:

$$\begin{vmatrix} \frac{\partial f}{\partial t}+u\frac{\partial f}{\partial x}+v\frac{\partial f}{\partial y}+w\frac{\partial f}{\partial z} & \rho\frac{\partial f}{\partial x} & \rho\frac{\partial f}{\partial y} & \rho\frac{\partial f}{\partial z} \\ \frac{a^2}{\rho}\frac{\partial f}{\partial x} & \frac{\partial f}{\partial t}+u\frac{\partial f}{\partial x}+v\frac{\partial f}{\partial y}+w\frac{\partial f}{\partial z} & 0 & 0 \\ \frac{a^2}{\rho}\frac{\partial f}{\partial y} & 0 & \frac{\partial f}{\partial t}+u\frac{\partial f}{\partial x}+v\frac{\partial f}{\partial y}+w\frac{\partial f}{\partial z} & 0 \\ \frac{a^2}{\rho}\frac{\partial f}{\partial z} & 0 & 0 & \frac{\partial f}{\partial t}+u\frac{\partial f}{\partial x}+v\frac{\partial f}{\partial y}+w\frac{\partial f}{\partial z} \end{vmatrix}=0.$$

This simplifies to

$$\left(\frac{Df}{Dt}\right)^2\left\{\left(\frac{Df}{Dt}\right)^2-a^2\operatorname{grad}^2 f\right\}=0, \quad (13.6.5)$$

where as usual $\operatorname{grad} f=\{\partial f/\partial x,\ \partial f/\partial y,\ \partial f/\partial z\}$ and following a practice which is very common in textbooks of classical hydrodynamics we have put

$$\frac{Df}{Dt}=\frac{\partial f}{\partial t}+u\frac{\partial f}{\partial x}+v\frac{\partial f}{\partial y}+w\frac{\partial f}{\partial z},$$

so that Df/Dt gives the rate of change of f measured at a point moving with the fluid. It can be seen at once from (13.6.5) that all the roots of this equation can give rise to real characteristic surfaces. The repeated root

$$\frac{Df}{Dt}=0 \quad (13.6.6)$$

represents a surface which remains stationary relative to the moving fluid, and so is a surface, or 'wave front' carried along by the fluid. On the other hand the surfaces satisfying the equations

$$\frac{Df}{Dt} = \pm a \,|\operatorname{grad} f| \tag{13.6.7}$$

are moving relative to the fluid in such a way that the component of their velocity in the direction of the instantaneous normal to the surface is the velocity of sound *a*.

APPENDIX 1

The Gauss Theorem in n Dimensions

As in some previous sections we shall prove the theorem in n dimensions, but rely on a diagram, Fig. A.1, which illustrates the three-dimensional case.

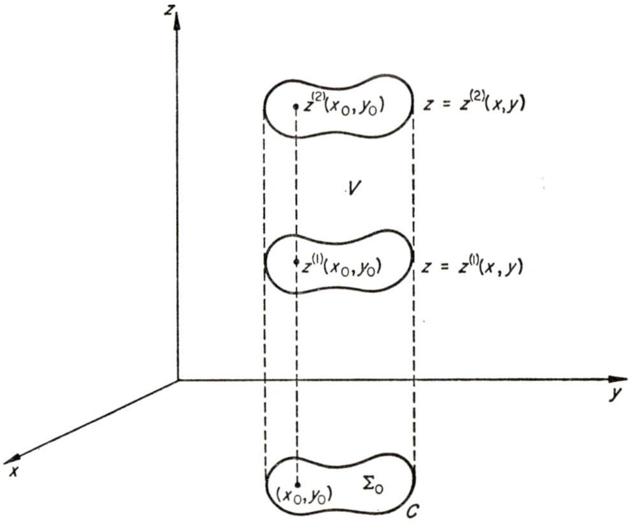

Fig. A.1

Let Σ_i be a closed, bounded region of the plane $x_i = 0$, bounded by C_i so that

$$\int_{\Sigma_i} d\tau_i > 0$$

where $d\tau_i = dx_1\, dx_2 \ldots dx_{i-1}\, dx_{i+1} \ldots dx_m$, and so that C_i has a tangent (hyperplane) almost everywhere. Let there be two surfaces

$$x_i^{(1)} = x_i^{(1)}\{x_1, x_2, \ldots, x_{i-1}, x_{i+1}, \ldots, x_m\}$$
$$x_i^{(2)} = x_i^{(2)}\{x_1, x_2, \ldots, x_{i-1}, x_{i+1}, \ldots, x_m\}$$

with the properties that

(*a*) for each point $\{x_1, x_2, \ldots, x_{i-1}, x_{i+1}, \ldots, x_m\}$ within or on the boundary of Σ_i,

$$x_i^{(1)} \leqslant x_i^{(2)};$$

(*b*) at each point of $x_i^{(1)}$ and $x_i^{(2)}$ whose projection on to $x_i = 0$ lies within or on the boundary of Σ_i, the curvature is finite and continuous;

(*c*) each line parallel to the axis of x_i, drawn through a point within or on the boundary of Σ_i, cuts $x_i^{(1)}$ in a single point and cuts $x_i^{(2)}$ in a single point.

We then can define a cylindrical region V bounded by a surface obtained by drawing through each point of C_i a line parallel to the x_i axis, and the intercept which this makes on the two surfaces $x_i^{(1)}$ and $x_i^{(2)}$.

Now suppose that $X(x_1, x_2, \ldots, x_{i-1}, x_i, x_{i+1}, \ldots, x_m)$ is a function of $\mathbf{x}$, so that $\partial X/\partial x_i$ is a continuous function of its arguments in and on the boundary of V. Then if

$$d\tau = dx_1\, dx_2 \ldots dx_{i-1}\, dx_i\, dx_{i+1} \ldots dx_m = d\tau_i\, dx_i$$

$$\int_V \frac{\partial X}{\partial x_i} d\tau = \int_{\Sigma_i} d\tau_i \int_{x_i^{(1)}}^{x_i^{(2)}} \frac{dX}{\partial x_i} dx_i = \int_{\Sigma_i} \{X^{(2)} - X^{(1)}\}\, d\tau_i.$$

We have put $X^{(j)} = X\{x_1, x_2, \ldots, x_{i-1}, x_i^{(j)}, x_{i+1}, \ldots, x_m\}$.

If $dS^{(1)}$ and $dS^{(2)}$ are increments of surface on $x_i^{(1)}$ and $x_i^{(2)}$, then (without summing the suffixes), on $x_i^{(1)}$

$$d\tau_i = -n_i^{(1)}\, dS^{(1)},$$

and on $x_i^{(2)}$

$$d\tau_i = n_i^{(2)}\, dS^{(2)},$$

where $n_i^{(j)}$ are the ith components of the unit vector normal to the surfaces $x_i^{(j)}$, and outward directed from the volume.

Hence we can write this as

$$\int_V \frac{dX}{\partial x_i} d\tau = \int_{S^1+S^2} n_i X\, dS,$$

where S^1 and S^2 are those parts of $x_i^{(1)}$ and $x_i^{(2)}$ which form a boundary of V.

It therefore follows that if V is any bounded volume of space in m dimensions, which can be broken up into a finite number of such cylindrical regions, we may add these together and so

$$\int_V \frac{\partial X}{\partial x_i} d\tau = \int_S n_i X\, dS,$$

where S is the surface bounding V, and n is the ith component of the outward normal to S. We may also clearly derive similar results with the i exchanged for any other suffix.

If then

(1) V is a bounded volume of space, with boundary S which has finite and continuous curvature at every point, except possibly at a finite set of loci on S of total surface measure zero,

(2) if, for each suffix i, $\partial X_i/\partial x_i$ is a function continuous in and on the boundary of V, except possibly at a finite number of intersecting hypersurfaces with properties similar to the boundary S, then

$$\int_V \left\{ \sum_{i=1}^{m} \frac{\partial X_i}{\partial x_i} \right\} d\tau = \int_S \left\{ \sum_{i=1}^{m} n_i\, X_i \right\} dS.$$

APPENDIX 2

The Volume and Surface Area of the n-dimensional Unit Sphere

The volume is simply expressed as

$$v_n = \int_{V_n} d\tau$$

where V_n is the region of space

$$x_1{}^2 + x_2{}^2 + \ldots + x_n{}^2 \leqslant 1.$$

If $0 < \xi < 1$ the hyperplane $x_n = \xi$ intersects the sphere in

$$x_1{}^2 + x_2{}^2 + \ldots + x_{n-1}{}^2 \leqslant 1 - \xi^2,$$

and the intersection with $x_n = -\xi$ has a similar equation. Thus the $(n-1)$-dimensional volume of either intersection is

$$v_{n-1}(1 - \xi^2)^{(n-1)/2}.$$

Hence

$$v_n = v_{n-1}\int_{-1}^{1}(1 - \xi^2)^{(n-1)/2}\,d\xi.$$

Now if $\eta = \xi^2$

$$\int_{-1}^{1}(1 - \xi^2)^{(n-1)/2}\,d\xi = \int_{0}^{1}\eta^{-1/2}(1 - \eta)^{(n-1)/2}\,d\eta = B\left\{\tfrac{1}{2}, \frac{n+1}{2}\right\}.$$

Thus

$$\Gamma\left\{\frac{n+2}{2}\right\}v_n = \sqrt{\pi}\,\Gamma\left\{\frac{n+1}{2}\right\}v_{n-1},$$

and so

$$\Gamma\left\{\frac{n+2}{2}\right\}v_n = C\pi^{n/2}$$

for some constant C. The volume of the unit dimensional sphere is, however, the length of the line segment between -1 and $+1$, and is therefore 2.

Thus

$$\Gamma\left\{\frac{n+2}{2}\right\}v_n = \pi^{n/2}.$$

The volume of the sphere of radius r in n dimensions is therefore

$$(\sqrt{\pi}r)^n/\Gamma\left(\frac{n+2}{2}\right).$$

The rate of increase of this volume with r is equal to the surface area of the sphere, that is if Ω_n is the surface area of the unit sphere in n dimensions,

$$\frac{d}{dr}\left\{(\sqrt{\pi}r)^n/\Gamma\left(\frac{n+2}{2}\right)\right\} = \Omega_n r^{n-1}.$$

Hence

$$\Omega_n = \left\{n(\sqrt{\pi})^n/\Gamma\left(\frac{n+2}{2}\right)\right\} = 2\pi^{n/2}/\Gamma\left(\frac{n}{2}\right).$$

APPENDIX 3

The Expansion Theorems of Fourier

DEFINITIONS

(*a*) A series of the form

$$\frac{a_0}{2} + \sum_{n=1}^{\infty} \{a_n \cos nx + b_n \sin nx\}$$

is called a trigonometrical series.

(*b*) If for some integer N, $a_n = b_n = 0$ when $n > N$, the series is said to represent a trigonometrical polynomial.

(*c*) If for some function $f(x)$ for which the integrals all exist,

$$\pi a_n = \int_{\alpha}^{\alpha+2\pi} f(\xi) \cos n\xi \, d\xi \quad \text{and} \quad \pi b_n = \int_{\alpha}^{\alpha+2\pi} f(\xi) \sin n\xi \, d\xi,$$

the series is said to be the Fourier series of $f(x)$ in the interval

$$\alpha < x < \alpha + 2\pi,$$

and a_n and b_n are said to be Fourier coefficients of $f(x)$.

(*d*) If for some $f(x)$

$$\pi a_n = 2\int_{\alpha}^{\alpha+\pi} f(\xi) \cos n\xi \, d\xi \quad \text{and} \quad b_n = 0,$$

the series is said to be the Fourier cosine series for $f(x)$ in

$$\alpha < x < \alpha + \pi.$$

There is an obvious and corresponding definition for the Fourier sine series.

Since we may always introduce a simple change of origin into the definition of the variable x, there is no loss of generality in taking $\alpha = -\pi$ in the definition of the Fourier series, and $\alpha = 0$ in the definition of the cosine and sine series. By simple scaling we may also change the interval $(-\pi, \pi)$ to any other finite interval $(-a, a)$.

We further define $\hat{f}(x)$ to be the 'periodic extension' of $f(x)$ out of $[-\pi, \pi]$; that is $\hat{f}(x)$ is a periodic function of period 2π which in $[-\pi, \pi]$ coincides with $f(x)$. Finally we put

$$2f^*(x) = \hat{f}(x-0) + \hat{f}(x+0) = \lim_{\varepsilon \to 0} \{ \hat{f}(x-\varepsilon) + \hat{f}(x+\varepsilon)\}.$$

THEOREM 1

If $f(x)$ is of bounded total variation in $-\pi \leqslant x \leqslant \pi$ and if

$$\int_{-\pi}^{\pi} |f(x)|\, dx < \infty,$$

then the Fourier series for $f(x)$ converges to $f^*(x)$.

We require two lemmas.

LEMMA 1 (the Riemann-Lebesgue lemma). If $f(x)$ is of bounded total variation in $[a, b]$ then if

$$C_\lambda(f) = \int_a^b f(x) \cos \lambda x \, dx \quad \text{and} \quad S_\lambda(f) = \int_a^b f(x) \sin \lambda x \, dx$$

then $C_\lambda(f) \to 0$ and $S_\lambda(f) \to 0$ as $\lambda \to \infty$.

Since $f(x)$ is of bounded total variation in $[a, b]$ it is bounded in the interval, and is also Riemann integrable in the interval. Hence, given ε, we can find a dissection of the interval

$$a = x_0 < x_1 < \ldots < x_n = b,$$

so that if $\delta_i = x_i - x_{i-1}$, and if M_i and m_i are respectively the maximum and minimum of $f(x)$ in $x_{i-1} < x \leqslant x_i$ then

$$\sum_{i=1}^{n} (M_i - m_i)\delta_i < \frac{\varepsilon}{2}.$$

In each interval choose a constant c_i so that $m_i \leqslant c_i \leqslant M_i$, and let $g(x)$ be the step function wherein

$$g(a) = c_1, \quad \text{and} \quad g(x) = c_i \quad \text{for} \quad x_{i-1} < x \leqslant x_i.$$

Then

$$C_\lambda(f) - C_\lambda(g) = \sum_{i=1}^{n} \int_{x_{i-1}}^{x_i} \{f(x) - c_i\} \cos \lambda x \, dx,$$

and so

$$| C_\lambda(f) - C_\lambda(g) | \leqslant \sum_{i=1}^{n} \{M_i - m_i\}\, \delta_i < \frac{\varepsilon}{2}.$$

On the other hand

$$C_\lambda(g) = \sum_{i=1}^{n} c_i \int_{xi_{-1}}^{xi} \cos \lambda x \, dx = \lambda^{-1} \sum_{i=1}^{n} c_i \{\sin(\lambda x_i) - \sin(\lambda x_{i-1})\},$$

and

$$| C_\lambda(g) | \leqslant 2\lambda^{-1} \sum_{i=1}^{n} | c_i |.$$

The last sum is finite, since $f(x)$ is of bounded variation, and so we may find a λ_0 such that if $\lambda > \lambda_0$

$$| C_\lambda(g) | < \frac{\varepsilon}{2}.$$

Finally therefore

$$| C_\lambda(f) | \leqslant | C_\lambda(f) - C_\lambda(g) | - | C_\lambda(g) | < \varepsilon.$$

A similar result may be proved for $S_\lambda(f)$.

LEMMA 2 (Dirichlet's integral). If

$$S_N = \frac{a_0}{2} + \sum_{n=1}^{N} (a_n \cos nx + b_n \sin nx)$$

where for each n, a_n and b_n are Fourier coefficients of a function $f(x)$ satisfying the restrictions of the main theorem, and if S is any constant, then

$$2\pi(S_N - S) = \int_0^\pi \phi(t) \sin \{(N + \tfrac{1}{2})t\} \operatorname{cosec} \left\{\frac{t}{2}\right\} dt,$$

wherein

$$\phi(t) = \hat{f}(x + t) + \hat{f}(x - t) - 2S.$$

From the definition of a_n and b_n

$$\pi\{a_n \cos nx + b_n \sin nx\} = \int_{-\pi}^{\pi} f(\xi)\{\cos nx \cos n\xi + \sin nx \sin n\xi\}\, d\xi,$$

$$= \int_{-\pi}^{\pi} f(\xi) \cos \{n(x - \xi)\}\, d\xi,$$

$$= \int_{-\pi-x}^{\pi-x} f(x + t) \cos nt\, dt,$$

if we make the substitution $\xi = x + t$. Now if $\pi - x \leqslant t \leqslant -\pi - x$, the argument of $f(x + t)$ lies between $-\pi$ and π, and we may replace $f(x + t)$ by $\hat{f}(x + t)$ in this integral. This is a periodic function, however, as is also $\cos nt$, and so its integral over a full period 2π is the same, regardless of the initial point of the interval. Thus

$$\pi(a_n \cos nx + b_n \sin nx) = \int_{-\pi}^{\pi} \hat{f}(x + t) \cos nt\, dt.$$

Summing this from $n = 0$ to $n = N$ we get

$$2\pi S_N = \int_{-\pi}^{\pi} f(x+t)\{1 + 2\cos t + 2\cos 2t + \ldots + 2\cos Nt\}\,dt,$$

$$= \int_{-\pi}^{\pi} f(x+t) \sin\{(N+\tfrac{1}{2})t\} \operatorname{cosec}\left\{\frac{t}{2}\right\} dt,$$

$$= \int_{0}^{\pi} \{f(x+t) + f(x-t)\} \sin\{(N+\tfrac{1}{2})t\} \operatorname{cosec}\left\{\frac{t}{2}\right\} dt.$$

This is the desired result with $S = 0$. If we replace $f(x)$ by $f(x) - S$, we must also replace $f(x+t) + f(x-t)$ by $\phi(t)$. On the other hand the only Fourier coefficient which is changed is a_0, which is replaced by $a_0 - 2S$. This changes S_N to $S_N - S$ and we derive the result stated.

Now put $S = f^*(x)$ into the Dirichlet integral, so that we get

$$2\pi\{S_N - f^*(x)\} = R_N^+ + R_N^-,$$

where we put

$$R_N^+ = \int_0^{\pi} \{f(x+t) - f(x+0)\} \sin\{(N+\tfrac{1}{2})t\} \operatorname{cosec}\left\{\frac{t}{2}\right\} dt,$$

and

$$R_N^- = \int_0^{\pi} \{f(x-t) - f(x-0)\} \sin\{(N+\tfrac{1}{2})t\} \operatorname{cosec}\left\{\frac{t}{2}\right\} dt.$$

Consider R_N^+ in detail. It is convenient to put $t = 2\theta$. Then $f(x+2\theta) - f(x+0)$, considered as a function of θ, is of bounded total variation in $0 \leqslant \theta \leqslant \pi/2$, and it tends to zero with θ. Also $\theta^{-1}\sin\theta$ is of bounded total variation in $0 \leqslant \theta \leqslant \pi/2$; it approaches unity as θ approaches zero, and is bounded below. Hence

$$\theta \operatorname{cosec} \theta\{f(x+2\theta) - f(x+0)\}$$

is of bounded total variation in $0 \leqslant \theta \leqslant \pi/2$, and vanishes with θ. By the usual theorem then

$$\theta \operatorname{cosec} \theta\{f(x+2\theta) - f(x+0)\} = \chi_1(\theta) - \chi_2(\theta),$$

in $0 \leqslant \theta \leqslant \pi/2$, where both $\chi_1(\theta)$ and $\chi_2(\theta)$ are monotonic functions (which we can assume without loss of generality to be positive increasing) and both vanish with θ.

We can, for given ε, find a δ such that if $0 \leqslant \theta < 2\delta$, $0 \leqslant \chi_1(\theta) < \varepsilon$ and $0 \leqslant \chi_2(\theta) < \varepsilon$. Thus

$$R_N^+ = \int_{\delta}^{\pi/2} \{f(x+2\theta) - f(x+0)\} \sin\{(2N+1)\theta\} \operatorname{cosec}\theta\, d\theta$$

$$+ \int_0^{\delta} \theta^{-1}\chi_1(\theta) \sin\{(2N+1)\theta\}\, d\theta$$

$$- \int_0^{\delta} \theta^{-1}\chi_2(\theta) \sin\{(2N+1)\theta\}\, d\theta.$$

Now

$$\left| \int_0^\delta \theta^{-1}\chi_1(\theta) \sin \{(2N+1)\theta\}\, d\theta \right| \leqslant \left| \chi_1(\delta) \int_{\theta_0}^\delta \theta^{-1} \sin \{(2N+1)\theta\}\, d\theta \right|,$$

for some $0 \leqslant \theta_0 \leqslant \delta$ by the mean value theorem. Furthermore, if A is the maximum of the bounded function

$$\left| \int_x^\infty t^{-1} \sin t\, dt \right|,$$

in $0 \leqslant x < \infty$, the right-hand side of the inequality above is bounded by $2A\varepsilon$. By an exactly similar argument

$$\left| \int_0^\delta \theta^{-1}\chi_2(\theta) \sin \{(2N+1)\theta\}\, d\theta \right| \leqslant 2A\varepsilon.$$

By the Riemann-Lebesgue lemma, given ε we can choose N so large that

$$\left| \int_\delta^{\pi/2} \{f(x+2\theta) - f(x+0)\} \sin \{(2N+1)\theta\} \operatorname{cosec} \theta\, d\theta \right| < \varepsilon.$$

Finally, therefore,

$$|R_N^+| < (4A+1)\varepsilon.$$

An exactly similar argument can be applied to R_N^-, and so as $N \to \infty$, S_N converges to $f^*(x)$.

We can obtain a stronger result than this. A series is said to be summable (C, 1) to S if

$$\frac{1}{N}\sum_{r=0}^{N-1} S_r \to S,$$

as $N \to \infty$, where S_r are the usual partial sums. Then S is said to be the first Cesàro sum of the series. We may then replace Theorem 1 by

THEOREM 2 (Fejér's theorem)

If $\int_{-\pi}^{\pi} f(x)\, dx$ exists and the integral converges absolutely (that is also $\int_{-\pi}^{\pi} |f(x)|\, dx < \infty$) the Fourier series converges (C, 1) to $f^*(x)$ at every point at which this is defined.

The proof is remarkably similar. We notice, however, that the only place in the proof of Theorem 1 where the property of bounded total variation is essential is where we divide part of the integrand into $\chi_1(\theta) - \chi_2(\theta)$. The significance of this is that to apply the mean value theorem we need the non-negative property of $\chi_1(\theta)$ and $\chi_2(\theta)$. Now using the Dirichlet theory above:

$$\pi \sum_{r=0}^{N-1} \{S_r - f^*(x)\} = \int_0^\pi \phi(t) \operatorname{cosec} \frac{t}{2} \left[\sum_{r=0}^{N-1} \sin \{(r + \tfrac{1}{2})t\} \right] dt$$

$$= \int_0^\pi \phi(t) \sin^2 \left\{ \frac{Nt}{2} \right\} \operatorname{cosec}^2 \left\{ \frac{t}{2} \right\} dt,$$

and the positive character of $\sin^2\{Nt/2\} \operatorname{cosec}^2\{t/2\}$ is sufficient for the mean value theorem. Thus we may replace the essential step by

$$\frac{1}{N} \left| \int_0^\delta \phi(t) \sin^2 \left\{ \frac{Nt}{2} \right\} \operatorname{cosec}^2 \left\{ \frac{t}{2} \right\} dt \right|$$

$$\leqslant \frac{\varepsilon}{N} \int_0^\pi \sin^2 \left\{ \frac{Nt}{2} \right\} \operatorname{cosec}^2 \left\{ \frac{t}{2} \right\} dt$$

$$\leqslant \frac{\varepsilon}{N} \int_0^\pi \{N + 2(N-1) \cos t + 2(N-2) \cos 2t + \dots$$

$$+ 2 \cos (N-1)t\} \, dt$$

$$\leqslant \pi\varepsilon.$$

If the Fourier series converges, at any point of continuity of $f(x)$, it converges to $f(x)$.

THEOREM 3

If

$$\int_{-\pi}^{\pi} | f(x) |^2 \, dx$$

exists, and the Fourier coefficients of $f(x)$ exist, then the integral

$$I_N = \int_{-\pi}^{\pi} \left\{ f(x) - (a_0/2) - \sum_{n=1}^{N} (a_n \cos nx + b_n \sin nx) \right\}^2 dx$$

is minimized if a_n and b_n are the Fourier coefficients.

$$\frac{\partial I_N}{\partial a_0} = \frac{1}{2} \int_{-\pi}^{\pi} f(x) \, dx - \frac{\pi}{2} a_0 + \sum_{n=1}^{N} \left(a_n \int_{-\pi}^{\pi} \cos nx \, dx + b_n \int_{-\pi}^{\pi} \sin nx \, dx \right),$$

$$= \tfrac{1}{2} \left\{ \int_{-\pi}^{\pi} f(x) \, dx - \pi a_0 \right\}.$$

Similarly

$$\frac{\partial I_N}{\partial a_m} = \int_{-\pi}^{\pi} f(x) \cos mx \, dx - a_0/2 \int_{-\pi}^{\pi} \cos mx \, dx$$

$$- \sum_{n=1}^{N} \left(a_n \int_{-\pi}^{\pi} \cos nx \cos mx \, dx + b_n \int_{-\pi}^{\pi} \cos mx \sin nx \, dx \right)$$

$$= \int_{-\pi}^{\pi} f(x) \cos mx \, dx - \pi a_m,$$

with a similar result for b_m. Thus equating to zero $\partial I_N/\partial a_m$ and $\partial I_N/\partial b_m$ we see that I_N is stationary when a_n and b_n are Fourier coefficients, and it is clear that it can only be a minimum.

Since the integrand in I_N is a perfect square we see that

$$I_N \geqslant 0,$$

so that if a_n and b_n are Fourier coefficients

$$\int_{-\pi}^{\pi} \{f(x)\}^2\, dx \geqslant a_0 \int_{-\pi}^{\pi} f(x)\, dx + 2\sum_{n=1}^{N} \left\{a_n \int_{-\pi}^{\pi} f(x) \cos nx\, dx + b_n \int_{-\pi}^{\pi} f(x) \sin nx\, dx\right\}$$
$$- \int_{-\pi}^{\pi} \left\{(a_0/2) + \sum_{n=1}^{N} (a_n \cos nx + b_n \sin x)\right\}^2 dx$$
$$= \pi\left\{(a_0{}^2/4) + \sum_{n=1}^{N} (a_n{}^2 + b_n{}^2)\right\}.$$

This is the Bessel inequality. Exactly similar results to both Theorem 3 and this inequality hold for other complete orthogonal systems of functions.

The left-hand side of this inequality is fixed, while the right-hand side increases with N. Thus if we replace N by infinity, the series on the right is convergent. If in the limit

$$\int_{-\pi}^{\pi} \{f(x)\}^2\, dx = \pi\left\{(a_0{}^2/4) + \sum_{n=1}^{\infty} (a_n{}^2 + b_n{}^2)\right\},$$

we say that the Fourier series converges in mean square to $f(x)$. It is obvious from the definition of I_N that a Fourier series which converges, converges in mean square, but the converse is not true.

We now extend Theorem 1 to the infinite interval.

THEOREM 4

If $f(x)$ is continuous on the whole line $-\infty < x < \infty$ except for, at most, a finite number of discontinuities, is of bounded total variation in $(-\infty, \infty)$ and

$$\int_{-\infty}^{\infty} |f(x)|\, dx < \infty,$$

then

$$\lim_{\lambda\to\infty} \int_{-\lambda}^{\lambda} d\alpha \int_{-\infty}^{\infty} \exp\{i\alpha(\xi - x)\} f(\xi)\, d\xi = \pi f^*(x)$$
$$= \tfrac{1}{2}\pi\{f(x+0) + f(x-0)\}.$$

We require an extension of Lemma 2.

Lemma 3. If in $-\infty < a \leqslant x \leqslant b < \infty$, $f(x)$ is of bounded total variation and has at most a finite number of discontinuities in $[a, b]$ then

$$\lim_{\lambda\to\infty} \int_a^b \frac{\sin\{\lambda(x-\xi)\}}{x-\xi} f(\xi)\, dx = \frac{\pi}{2}\{f(x+0)+f(x-0)\}.$$

In fact $f(\xi)/(x-\xi)$ is of bounded total variation in both intervals $[a, x-\delta]$, and $[x+\delta, b]$ for every $\delta > 0$, so that by the Riemann-Lebesgue lemma these parts of the interval do not contribute to the limit of the integral on the left-hand side.

Thus we may replace the left-hand side by

$$\lim_{\lambda\to\infty} \int_{-\delta}^{\delta} \frac{\sin \lambda t}{t} f(x+t)\, dt = \lim_{\lambda\to\infty} \int_0^{\delta} \frac{\sin \lambda t}{t} \{f(x+t)+f(x-t)\}\, dt.$$

Put

$$\phi^+(t) = f(x+t) - f(x+0), \quad \phi^-(t) = f(x-t) - f(x-0)$$

and

$$\phi(t) = \phi^+(t) + \phi^-(t).$$

Both $\phi^+(t)$ and $\phi^-(t)$ are of bounded total variation in $0 \leqslant t \leqslant \delta$ and vanish at $t = 0$, and so we can represent each of them as the difference of two monotonic functions that we may assume positive. The argument now follows the same pattern as the end of the proof of Theorem 1. Thus

$$\lim_{\lambda\to\infty} \int_0^{\delta} \frac{\sin \lambda t}{t} \phi(t)\, dt = 0,$$

and so

$$\begin{aligned}\lim_{\lambda\to\infty} \int_a^b \frac{\sin\{\lambda(x-\xi)\}}{x-\xi} f(\xi)\, d\xi &= f^*(x) \lim_{\lambda\to\infty} \int_0^{\delta} \frac{\sin \lambda t}{t}\, dt \\ &= f^*(x) \lim_{\lambda\to\infty} \int_0^{\lambda\delta} \frac{\sin \tau}{\tau}\, d\tau, \\ &= \frac{\pi}{2} f^*(x).\end{aligned}$$

Furthermore

$$\left| \int_b^{\infty} \frac{\sin\{\lambda(x-\xi)\}}{x-\xi} f(\xi)\, d\xi \right| \leqslant \frac{1}{|b| - |x|} \int_b^{\infty} |f(\xi)|\, d\xi,$$

if $|b| > |x|$, and so as $b \to \infty$, if x is kept fixed, then

$$\left| \int_b^{\infty} \frac{\sin\{\lambda(x-\xi)\}}{x-\xi} f(\xi)\, d\xi \right| \to 0,$$

since

$$\int_{-\infty}^{\infty} |f(\xi)|\, d\xi < \infty$$

with a similar result for the other limit. This enables us to extend the lemma to read

$$\lim_{\lambda\to\infty}\int_{-\infty}^{\infty}\frac{\sin\{\lambda(x-\xi)\}}{x-\xi}f(\xi)\,d\xi=\frac{\pi}{2}f^*(x),$$

or, which is the same thing,

$$\lim_{\lambda\to\infty}\int_{-\infty}^{\infty}\left[\int_0^{\lambda}\cos\{\alpha(x-\xi)\}\,d\alpha\right]f(\xi)\,d\xi=\frac{\pi}{2}f^*(x).$$

Let

$$\Phi(\xi,\alpha)=f(\xi)\cos\{\alpha(x-\xi)\}.$$

$$\left|\int_0^{\infty}\left\{\int_0^{\lambda}\Phi(\xi,\alpha)\,d\alpha\right\}d\xi-\int_0^{\lambda}\left\{\int_0^{\infty}\Phi(\xi,\alpha)\,d\xi\right\}d\alpha\right|$$

$$=\left|\int_0^{b}\left\{\int_0^{\lambda}\Phi(\xi,\alpha)\,d\alpha\right\}d\xi-\int_0^{\lambda}\left\{\int_0^{b}\Phi(\xi,\alpha)\,d\xi\right\}d\alpha\right.$$

$$\left.+\int_b^{\infty}\left\{\int_0^{\lambda}\Phi(\xi,\alpha)\,d\alpha\right\}d\xi-\int_0^{\lambda}\left\{\int_b^{\infty}\Phi(\xi,\alpha)\,d\xi\right\}d\alpha\right|,$$

$$=\left|\int_b^{\infty}\left\{\int_0^{\lambda}\Phi(\xi,\alpha)\,d\alpha\right\}d\xi-\int_0^{\lambda}\left\{\int_b^{\infty}\Phi(\xi,\alpha)\,d\xi\right\}d\alpha\right|$$

$$\leqslant\int_b^{\infty}\left\{\int_0^{\lambda}|\Phi(\xi,\alpha)|\,d\alpha\right\}d\xi+\int_0^{\lambda}\left\{\int_b^{\infty}|\Phi(\xi,\alpha)|\,d\xi\right\}d\alpha$$

$$=2\lambda\int_b^{\infty}|f(\xi)|\,d\xi.$$

Given $\varepsilon>0$ arbitrarily, for fixed λ we can choose b so large that this is less than ε. Hence

$$\int_0^{\infty}\left[\int_0^{\lambda}f(\xi)\cos\{\alpha(x-\xi)\}\,d\alpha\right]d\xi=\int_0^{\lambda}\left[\int_0^{\infty}f(\xi)\cos\{\alpha(x-\xi)\}\,d\xi\right]d\alpha,$$

and similarly

$$\int_{-\infty}^{0}\left[\int_0^{\lambda}f(\xi)\cos\{\alpha(x-\xi)\}\,d\alpha\right]d\xi=\int_0^{\lambda}\left[\int_{-\infty}^{0}f(\xi)\cos\{\alpha(x-\xi)\}\,d\xi\right]d\alpha.$$

Adding these we see that

$$\lim_{\lambda\to\infty}\int_{-\lambda}^{\lambda}\left[\int_{-\infty}^{\infty}f(\xi)\exp\{i\alpha(x-\xi)\}\,d\xi\right]d\alpha$$

$$=\lim_{\lambda\to\infty}2\int_0^{\lambda}\left[\int_{-\infty}^{\infty}f(\xi)\cos\{\alpha(x-\xi)\}\,d\xi\right]d\alpha,$$

$$=\lim_{\lambda\to\infty}2\int_{-\infty}^{\infty}\left[\int_0^{\lambda}f(\xi)\cos\{\alpha(x-\xi)\}\,d\alpha\right]d\xi=\pi f^*(x).$$

We can express this result more neatly in terms of the Fourier transform $\bar{f}(\alpha)$. For if

$$\bar{f}(\alpha) = (2\pi)^{-1/2}\int_{-\infty}^{\infty} f(\xi) \exp (i\alpha\xi)\, d\xi$$

then

$$\tfrac{1}{2}\{f(x + 0) + f(x - 0)\} = (2\pi)^{-1/2} \lim_{\lambda\to\infty} \int_{-\lambda}^{\lambda} \bar{f}(\alpha) \exp (-i\alpha x)\, d\alpha.$$

If $f(x)$ is a given function on $[0, \infty]$ with the properties required by Theorem 4, then we can continue it as an even function into the region $[-\infty, 0]$. In consequence we get the cosine transform

$$\bar{f}(\alpha) = \left(\frac{2}{\pi}\right)^{1/2}\int_{0}^{\infty} f(\xi) \cos \alpha\xi\, d\xi,$$

with

$$\tfrac{1}{2}\{f(x + 0) + f(x - 0)\} = \left(\frac{2}{\pi}\right)^{1/2}\int_{0}^{\infty} \bar{f}(\alpha) \cos \alpha\xi\, d\alpha.$$

If instead we continue it as an odd function we get the Fourier sine transform theorem. For a fully rigorous and extensive discussion of the convergence of Fourier series and integrals see [18], [20].

In spaces of dimensions greater than one, we may derive theorems corresponding to Theorems 1 and 4. In two dimensions, for example, by applying Theorem 4 twice successively, we may derive the following:

Theorem 5

If $f(x, y)$ is continuous on the whole (x, y) plane except possibly along a finite number of curves, is integrable on every closed domain, if $f(x, y)$ is of bounded variation in the whole plane, and if

$$\int_{-\infty}^{\infty}\int_{-\infty}^{\infty} | f(x, y) |\, dx\, dy < \infty,$$

and

$$\bar{f}(\alpha, \beta) = (2\pi)^{-1}\int_{-\infty}^{\infty}\int_{-\infty}^{\infty} f(x, y) \exp (i\alpha x + i\beta y)\, dx\, dy$$

then

$$(2\pi)^{-1}\lim_{\lambda\to\infty} \lim_{\mu\to\infty} \int_{-\lambda}^{\lambda}\int_{-\mu}^{\mu} \bar{f}(\alpha, \beta) \exp \{ - (i\alpha x + i\beta y)\}\, d\alpha\, d\beta$$

converges. At a point of continuity of $f(x, y)$ it converges to $f(x, y)$, but at a point on one of the curves of discontinuity it will converge to the mean of the values on either side of the discontinuity.

Suppose that $f(x, y) = f\{(x^2 + y^2)^{1/2}\}$, then this suggests that we introduce polar coordinates

$$x = r \cos \theta, \quad y = r \sin \theta,$$

and correspondingly we shall put

$$\alpha = \rho \cos \gamma, \quad \beta = \rho \sin \gamma.$$

Then

$$2\pi \bar{f}(\alpha, \beta) = \int_{-\pi}^{\pi} \int_{0}^{\infty} f(r) \exp \{i\rho r \cos (\theta - \gamma)\} r \, dr \, d\theta$$

$$= 2\pi \int_{0}^{\infty} f(r) r \, J_0(\rho r) \, dr,$$

from the integral definition of $J_0(\rho r)$ [22].

Hence the left-hand side is a function only of ρ and not of γ. Then if we may replace the limits $\lambda \to \infty$ and $\mu \to \infty$ by infinite limits, we see correspondingly that

$$f(r) = \int_{0}^{\infty} \rho \bar{f}(\rho) \, J_0(\rho r) \, d\rho.$$

If instead of putting $f(x, y) = f(r)$ we put $f(x, y) = f(r) \exp (-in\theta)$, and use the integral definition of $J_n(\rho r)$ [22] we can derive the reciprocal relation for the Hankel transform, that if

$$\bar{f}(\rho) = \int_{0}^{\infty} r f(r) \, J_n(\rho r) \, dr,$$

then

$$f(r) = \int_{0}^{\infty} \rho \bar{f}(\rho) \, J_n(\rho r) \, d\rho.$$

A similar reciprocal relation exists between the integral pair when $J_n(\rho r)$ is replaced by $J_\nu(\rho r)$ for any real $\nu > -\frac{1}{2}$. This may be established by a proof very little different from that used for the Fourier integral above, since, as was noted in Chapter 6, asymptotically the Bessel functions behave rather like the trigonometrical functions.

APPENDIX 4

Singularities of Legendre Functions

The Legendre differential equation of order ν,

$$(1 - z^2)\frac{d^2W}{dz^2} - 2z\frac{dW}{dz} + \nu(1 + \nu)W = 0,$$

is a hypergeometric equation. Thus it has three regular singular points, at $z = \pm 1$ and at $z = \infty$; every other point of the complex plane is ordinary.

Near the origin we can find two independent solutions by expansion in series

$$W_1(z) = 1 + \sum_{n=1}^{\infty} a_n z^{2n},$$

and

$$W_2(z) = z\left\{1 + \sum_{n=1}^{\infty} b_n z^{2n}\right\},$$

where

$$a_n = \frac{(2n-1+\nu)(2n-3+\nu)\ldots(1+\nu)(2n-2-\nu)(2n-4-\nu)\ldots(-\nu)}{\{(2n)!\}}$$

and

$$b_n = \frac{(2n+\nu)(2n-2+\nu)\ldots(2+\nu)(2n-1-\nu)(2n-3-\nu)\ldots(1-\nu)}{\{(2n+1)!\}}.$$

Now

$$\frac{a_{n+1}}{a_n} = \frac{(2n + 1 + \nu)(2n - \nu)}{(2n + 2)(2n + 1)}.$$

Hence the series defined by

$$1 + \sum_{n=1}^{\infty} a_n \zeta^n$$

has radius of convergence unity. Its coefficients are all real, if ν is real, and ultimately all of the same sign, since, regardless of ν, we can clearly find an n for which $2n - \nu > 0$. There is a well-known theorem which says that if a series expansion in powers of $z - a$ has all its coefficients real and of the same sign, and has finite radius of convergence R, then the function represented by the series has a singularity at the point $z = a + R$ (see [20]).

Therefore either this function has a singularity at $\zeta = 1$, or the series must terminate. This last can only happen if $\nu = 2N$ or $\nu = -(2N + 1)$ for some positive integer N.

Similarly we may consider

$$1 + \sum_{n=1}^{\infty} b_n \zeta^n,$$

and by an identical argument show that this must have a singularity at $\zeta = 1$, or the series will terminate, and this can only occur if $\nu = 2N + 1$ or $\nu = -2N$ for some positive integer N.

To sum up so far then: if ν is an even integer, $W_1(z)$ terminates, and $W_2(z)$ is an infinite series; if ν is an odd integer $W_2(z)$ terminates and $W_1(z)$ is an infinite series. If either series does terminate, it is a polynomial (differing only by a constant from the usually defined Legendre polynomial), and in consequence is an integral function. All other solutions will have singularities where $z^2 = 1$, that is at $z = \pm 1$.

Now look at the behaviour of solutions near $z = 1$. Both solutions of the indicial equation coincide and are zero, and so near $z = 1$ there are two independent solutions:

$$W_3(z) = 1 + \sum_{n=1}^{\infty} C_n (z - 1)^n,$$

and

$$W_4(z) = W_3(z) \log \{z - 1\} + \sum_{n=0}^{\infty} d_n (z - 1)^n.$$

The coefficients

$$C_n = \frac{\{(n-1)^2 - \nu(1+\nu)\}\{(n-2)^2 - \nu(1+\nu)\} \ldots \{-\nu(1+\nu)\}}{2^n \{n!\}^2}$$

and the exact form of the coefficients d_n are irrelevant to the argument. It can therefore be seen that $W_3(z)$ represents a solution regular in the circle $|z - 1| < 2$, while $W_4(z)$ has a logarithmic branch point at $z = 1$.

There is therefore a linear combination of $W_1(z)$ and $W_2(z)$ which is regular at $z = 1$, namely a multiple of $W_3(z)$. On the other hand

$W_3(z)$ must have a singularity on the circle $|z - 1| = 2$, and since every point on this circle, other than $z = -1$, is an ordinary point of the differential equation, this singularity must be at $z = -1$. By the symmetry of the equation in $z - 1$ and $z + 1$, this singularity will be logarithmic.

Finally, therefore, we conclude that if ν is an integer the Legendre equation has one solution which is an integral function, and its second solution will then have logarithmic singularities at $z = \pm 1$.

If ν is not an integer there are solutions regular at $z = 1$, and solutions regular at $z = -1$, but not both together.

In other words, the only solutions of the Legendre equation regular at $z = \pm 1$ are Legendre polynomials, and so this must require that ν is an integer.

For a fuller account of the properties of these solutions the reader should consult [24].

The proof above only needs the assumption that $\nu(\nu + 1)$ is real, not that ν itself is real. It can also be extended to apply to the equation for the associated Legendre functions.

APPENDIX 5

An Identity Involving Derivatives

Let
$$I_n(z) = \frac{d^n}{d(z^2)^n}\{z^{2n+1} f(z)\}. \tag{A.5.1}$$

Now if $f(0) = 0$, and $\mathscr{L}\{f(z)\} = \bar{f}(s)$ is the Laplace transform of $f(z)$, then

$$\mathscr{L}\left\{\frac{d}{d(z^2)} f(z)\right\} = \tfrac{1}{2}\int_0^\infty z^{-1} e^{-sz} \frac{df}{dz} dz = \tfrac{1}{2}\int_s^\infty \bar{f}(t) t\, dt, \tag{A.5.2}$$

$$\mathscr{L}\left\{\frac{d^2}{d(z^2)^2} f(z)\right\} = \tfrac{1}{4}\int_s^\infty dt\, t \int_t^\infty \bar{f}(u) u\, du = 2^{-4}\int_{s^2}^\infty (t^2 - s^2)\bar{f}(t)\, d(t^2),$$

if (A.5.3)

$$\left\{\frac{d}{d(z^2)} f(z)\right\}_{z=0} = 0.$$

Since in our case

$$\left[\frac{d^j}{d(z^2)^j}\{z^{2n+1} f(z)\}\right]_{z=0} = 0 \quad \text{for} \quad j = 0, 1, \ldots n-1,$$

we may deduce in a similar way that

$$\mathscr{L}\{I_n(z)\} = 2^{1-2n}(-1)^{2n+1}[(n-1)!]^{-1} \int_s^\infty t(t^2 - s^2)^{n-1} \frac{d^{2n+1}\bar{f}(t)}{dt^{2n+1}} dt. \tag{A.5.4}$$

Clearly $t(t^2 - s^2)^{n-1}$ is a polynomial in t of order $2n - 1$, so that (A.5.3) can be integrated by parts $2n$ times, which removes the integral sign. Thus

$$\mathscr{L}\{I_n(z)\} = 2^{1-2n}[(n-1)!]^{-1} \sum_{j=0}^{2n-1} (-1)^j \left[\frac{d^j}{dt^j} t(t^2 - s^2)^{n-1}\right]_{t=s} \frac{d^{2n-j}\bar{f}(s)}{ds^{2n-j}}$$

$$= 2^{1-2n}[(n-1)!]^{-1} \sum_{j=n-1}^{2n-1} (-1)^j \left[\frac{d^j}{dt^j} t(t^2 - s^2)^{n-1}\right]_{t=s} \frac{d^{2n-j}\bar{f}(s)}{ds^{2n-j}}. \tag{A.5.5}$$

Hence we may write

$$\mathscr{L}\{I_n(z)\} = \sum_{j=0}^{n} (-1)^j C_{n,j}\, s^{n-j} \frac{d^{n-j+1}\bar{f}(s)}{ds^{n-j+1}}, \tag{A.5.6}$$

where

$$C_{n,j} = 2^{1-2n} \sum_{k=0}^{n-1} \frac{(-1)^k\, \Gamma(2k+2)}{\Gamma(n-k)\, \Gamma(k+1)\, \Gamma(2k+3-j-n)}.$$

If the transform is inverted term by term

$$I_n(z) = (-1)^{n+1} \sum_{j=0}^{n} C_{n,j} \frac{d^{n-j}}{dz^{n-j}} \{z^{n-j+1} f(z)\}. \tag{A.5.7}$$

From above

$$C_{n,j} = 2^{j-n} \sum_{k=0}^{n-1} \frac{(-1)^k\, \Gamma(k+\frac{3}{2})}{\Gamma(n-k)\, \Gamma\{k+\frac{1}{2}(3-j-n)\}\, \Gamma\{k+2-\frac{1}{2}(j+n)\}} \tag{A.5.8}$$

and there are two cases to consider if $n + j = 2m$.

$$\begin{aligned} C_{n,j} &= 2^{2m-2n} \sum_{k=0}^{n-1} \frac{(-1)^k\, \Gamma(k+\frac{3}{2})}{\Gamma(n-k)\, \Gamma(k-m+\frac{3}{2})\, \Gamma(k-m+2)} \\ &= 2^{2m-2n} \sum_{k=m-1}^{n-1} \frac{(-1)^k\, \Gamma(k+\frac{3}{2})}{\Gamma(n-k)\, \Gamma(k-m+\frac{3}{2})\, \Gamma(k-m+2)} \\ &= 2^{2m-2n}(-1)^{m-1} \sum_{k=0}^{n-m} \frac{(-1)^k\, \Gamma(k+m+\frac{1}{2})}{\Gamma(n-m+1-k)\, \Gamma(k+\frac{1}{2})\, \Gamma(k+1)}. \end{aligned} \tag{A.5.9}$$

Now if $|t| < 1$

$$(1-t)^{n-m-1/2} = \sum_{r=0}^{\infty} (-1)^r \frac{\Gamma(n-m+\frac{1}{2})}{\Gamma(r+1)\, \Gamma(n-m-r+\frac{1}{2})} t^r$$

and

$$(1-t)^{-m-1/2} = \sum_{r=0}^{\infty} \frac{\Gamma(m+r+\frac{1}{2})}{\Gamma(r+1)\, \Gamma(m+\frac{1}{2})} t^r$$

and so

$$\begin{aligned} (1-t)^{n-2m-1} = (1-t)^{-j-1} &= \sum_{r=0}^{\infty} \frac{\Gamma(j+r+1) t^r}{\Gamma(r+1)\, \Gamma(j+1)} \\ &= (1-t)^{n-m-1/2}(1-t)^{-m-1/2} \\ &= \sum_{r=0}^{\infty} d_r t^r \end{aligned}$$

where

$$d_r = \sum_{k=0}^{r} (-1)^{r-k} \frac{\Gamma(n-m+\frac{1}{2})\,\Gamma(m+k+\frac{1}{2})}{\Gamma(r-k+1)\,\Gamma(n-m-r+k+\frac{1}{2})\,\Gamma(k+1)\,\Gamma(m+\frac{1}{2})}.$$

Thus

$$C_{n,j} = 2^{-2(n-m)}(-1)^{n-1}\frac{\Gamma(m+\frac{1}{2})}{\Gamma(n-m+\frac{1}{2})}d_{n-m}$$

$$= (-1)^{n-1}2^{-2(n-m)}\frac{\Gamma(m+\frac{1}{2})\,\Gamma(j+n-m+1)}{\Gamma(n-m+\frac{1}{2})\,\Gamma(n-m+1)\,\Gamma(j+1)}$$

$$= (-1)^{n-1}2^{-2(n-m)}\frac{\Gamma(m+\frac{1}{2})\,\Gamma(m+1)}{\Gamma(n-m+\frac{1}{2})\,\Gamma(n-m+1)\,\Gamma(j+1)}$$

$$= (-1)^{n-1}2^{-(n+j)}\frac{\Gamma(n+j+1)}{\Gamma(n-j+1)\,\Gamma(j+1)}.$$

Similarly if $n+j = 2m-1$

$$C_{n,j} = 2^{2m-2n-1}(-1)^{m-1}\sum_{k=0}^{n-m}\frac{(-1)^k\,\Gamma(k+m+\frac{1}{2})}{\Gamma(n-m-k+1)\,\Gamma(k+1)\,\Gamma(k+\frac{3}{2})}$$

$$= 2^{2m-2n-1}(-1)^{n-1}\frac{\Gamma(m+\frac{1}{2})}{\Gamma(n-m+\frac{3}{2})}D_{n-m}$$

where D_{n-m} is the coefficient of t^{n-m} in $(1-t)^{n-2m} = (1-t)^{-j-1}$. This once again leads to the same value for $C_{n,j}$.

References

[1] Courant, R. *Partial Differential Equations*, Interscience, New York (1962).

[2] Courant, R. and Friedrichs, K. O. *Supersonic Flow and Shock Waves*, Interscience, New York (1948).

[3] Ferrar, W. L. *Algebra*, Clarendon, Oxford (1957).

[4] Garabedian, P. R. *Partial Differential Equations*, Wiley, New York (1964).

[5] Gel'fand, I. M. and Shilov, G. E. *Generalised Functions*, Vol. **1**, Properties and Operations, Academic Press, New York and London (1964).

[6] Gibson, G. A. *Advanced Calculus*, Macmillan, London (1931).

[7] Goursat, E. *Differential Equations*, Ginn, Boston, U.S.A. (1945).

[8] Hadamard, J. *Lectures on Cauchy's Problem in Linear Partial Differential Equations*, Dover, New York (1952).

[9] Hörmander, L. *Linear Partial Differential Equations*, Springer, Berlin (1963).

[10] Howarth, L. (editor). *Modern Developments in Fluid Dynamics*, Clarendon, Oxford (1953).

[11] Ince, E. L. *Ordinary Differential Equations*, Dover, New York (1956).

[12] Lighthill, M. J. *An Introduction to Fourier Analysis and Generalised Functions*, Cambridge University Press, London (1958).

[13] Pai, S. *Magnetohydrodynamics and Plasma Physics*, Springer, Vienna (1962).

[14] Schwartz, L. *Théorie des distributions*, Vols **1** and **2**, Hermann, Paris (1957–59).

[15] Sneddon, I. N. *Fourier Transforms*, McGraw-Hill, New York (1951).

[16] Spain, B. *Vector Analysis*, Van Nostrand, London (1965).

[17] Temple, G. *Proc. R. Soc.*, **A228**, 175–90 (1955).

[18] Titchmarsh, E. C. *An Introduction to the Theory of Fourier Integrals*, Clarendon, Oxford (1948).

[19] Titchmarsh, E. C. *Eigenfunction Expansions associated with Second Order Differential Equations*, Vol. **1**, Clarendon, Oxford (1962).

[20] Titchmarsh, E. C. *The Theory of Functions*, Clarendon, Oxford (1939).

[21] Ward, G. N. *Linearised Theory of Steady High Speed Flows*, Cambridge University Press, London (1955).

[22] Watson, G. N. *Theory of Bessel Functions*, Cambridge University Press, London (1922).

[23] Webster, A. G. *Partial Differential Equations of Mathematical Physics*, Hafner, New York (1950).

[24] Whittaker, E. T. and Watson, G. N. *Modern Analysis*, (4th edn), Cambridge University Press, London (1952).

Index